& Physiology Handbook

Sue Seif, MA, CMI
Robert Edwards, MS, CMI

Jennifer Webb, MS
Cheryl Reynon, MA

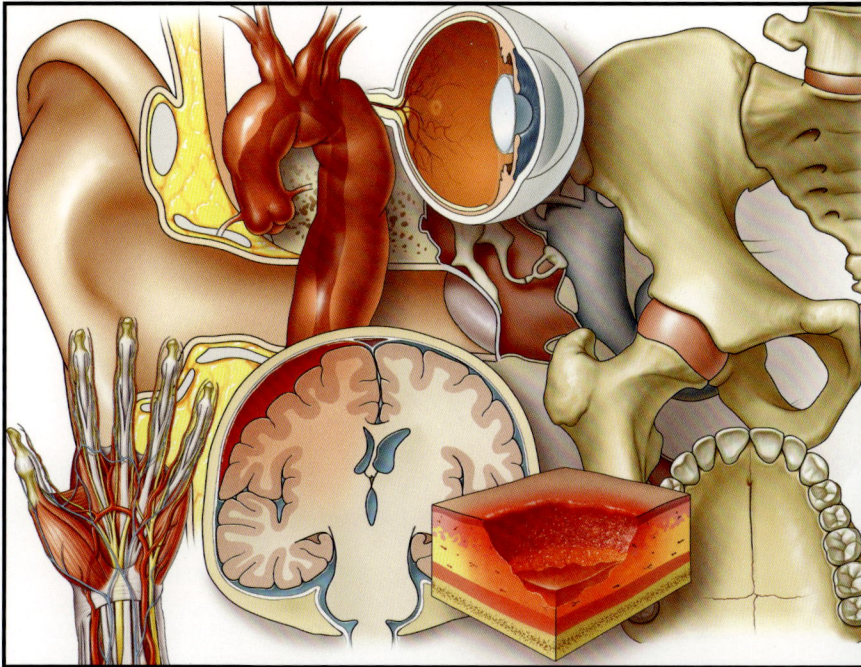

SEIF & ASSOCIATES INC.

M E D I C A L · G R A P H I C S

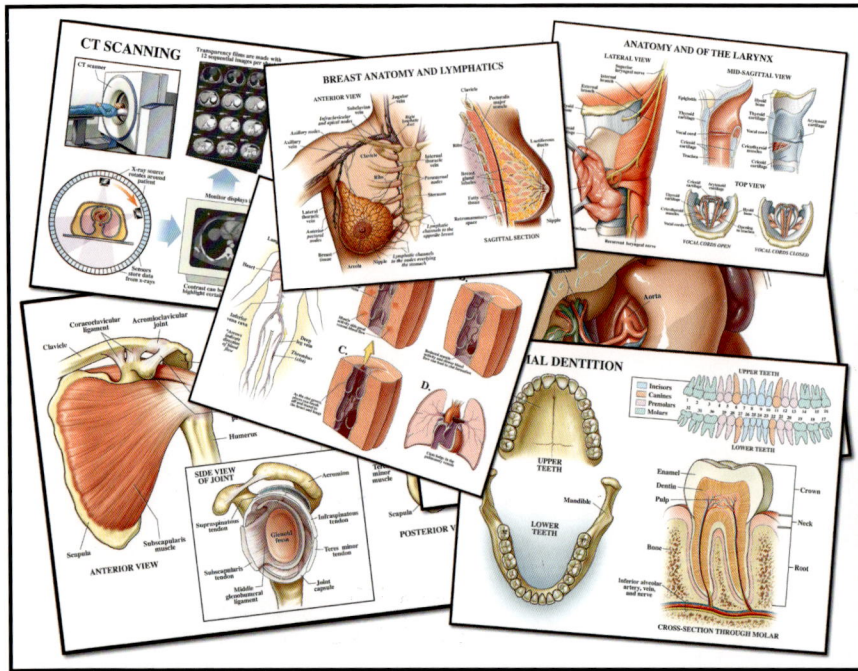

Introduction

This handbook is meant to be a thinking-person's guide to common medical issues. Its broad-based subject matter is designed to give an overview of the medicine behind the kinds of subjects we read about or deal with every day. What is a stroke, really? A heart attack? What happens during gallbladder surgery? Labor?

The images and text in this volume may be able to give the reader more insight into the workings of the body, and what kind of questions to ask his/her health care provider.

Everyone is different, and while basic anatomy is similar from person to person, there are many normal variations and many different ways for the same condition to express itself. No book can ever substitute for professional medical care!

CONTENTS

CONTENTS

NORMAL HEART ANATOMY

Common carotid arteries

Right subclavian artery

Left subclavian artery

Ascending aorta

Descending aorta

Pulmonary trunk

Superior vena cava

Left coronary artery

Right coronary artery

Left atrium

Circumflex branch

Left anterior descending artery (LAD)

Right atrium

Left ventricle

Right ventricle

ANTERIOR VIEW

Right pulmonary artery

Pulmonary veins

Inferior vena cava

Posterior descending artery

Right coronary artery

POSTERIOR VIEW

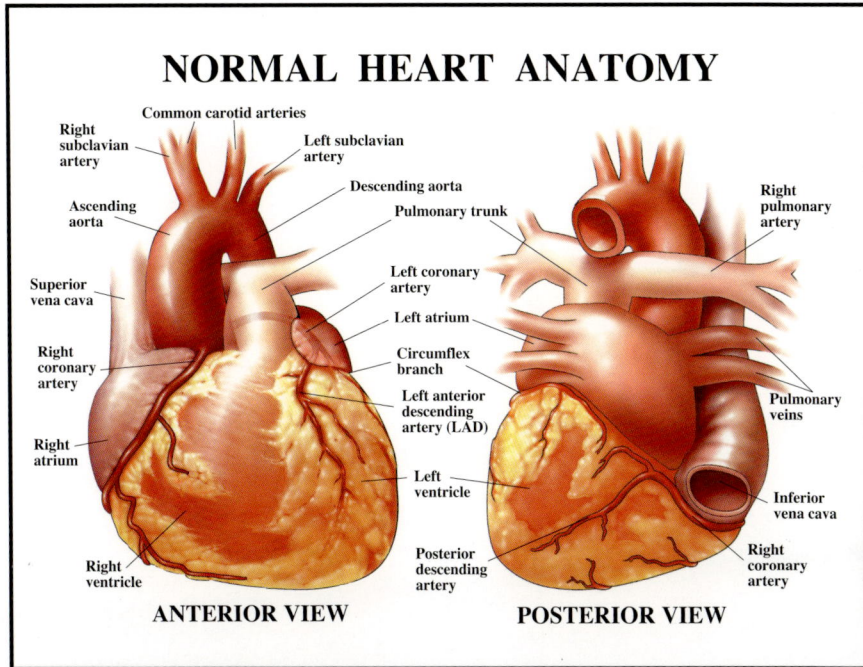

C1: Normal heart anatomy

- Heart muscle is supplied by the cornary arteries, not by the blood flowing through the heart.

- The major coronary vessels are the right coronary artery (RCA) and left main coronary artery (LCA), both of which come directly off of the aorta via the coronary ostia.

- The LCA divides into the left anterior descending artery(LAD) and circumflex artery.

- The RCA has no major branches, and terminates as the posterior descending artery (PDA).

- There are may be variations in the anatomy.

Note: "Right" and "left" always refer to the patient's right or left, not the viewer's.

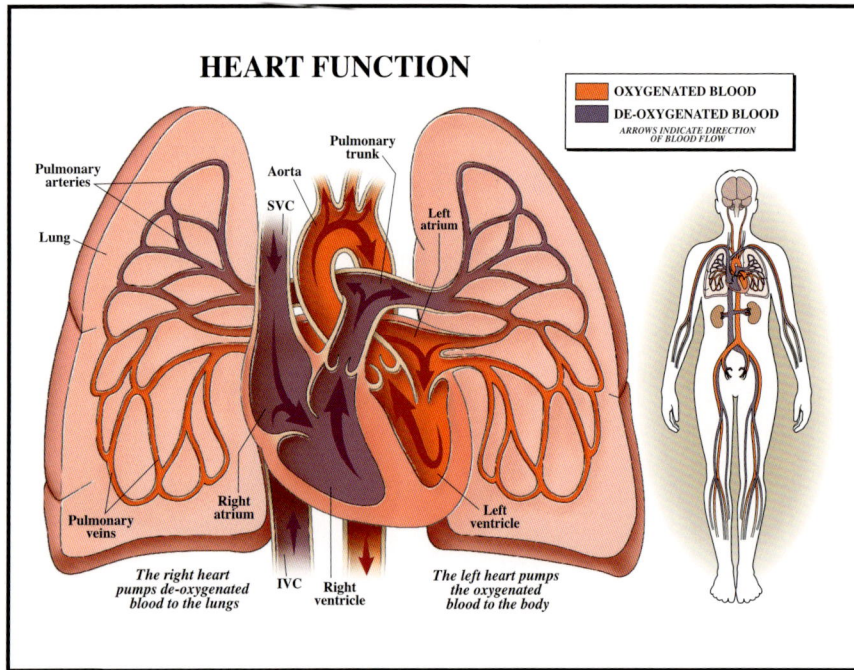

HEART FUNCTION

OXYGENATED BLOOD
DE-OXYGENATED BLOOD
ARROWS INDICATE DIRECTION OF BLOOD FLOW

Pulmonary arteries
Lung
Aorta
SVC
Pulmonary trunk
Left atrium
Pulmonary veins
Right atrium
IVC
Right ventricle
Left ventricle

The right heart pumps de-oxygenated blood to the lungs

The left heart pumps the oxygenated blood to the body

C2: Heart function

- The normal heart is really two separate pumps working in tandem; there is no connection between the right and left sides in the normal post-fetal heart.

- The right heart receives de-oxygenated blood from the body, moving it from the right atrium to the right ventricle to the lungs via the pulmonary artery. Carbon dioxide is released and oxygen is picked up in the lungs.

- The left heart receives oxygenated blood from the lungs, moving it from the left atrium to the left ventricle, and from there to the aorta, which distributes it to the rest of the body.

- The ventricles are thick muscular chambers which move blood with each contraction; the average left ventricle contracts with a force of 120 mmHg.

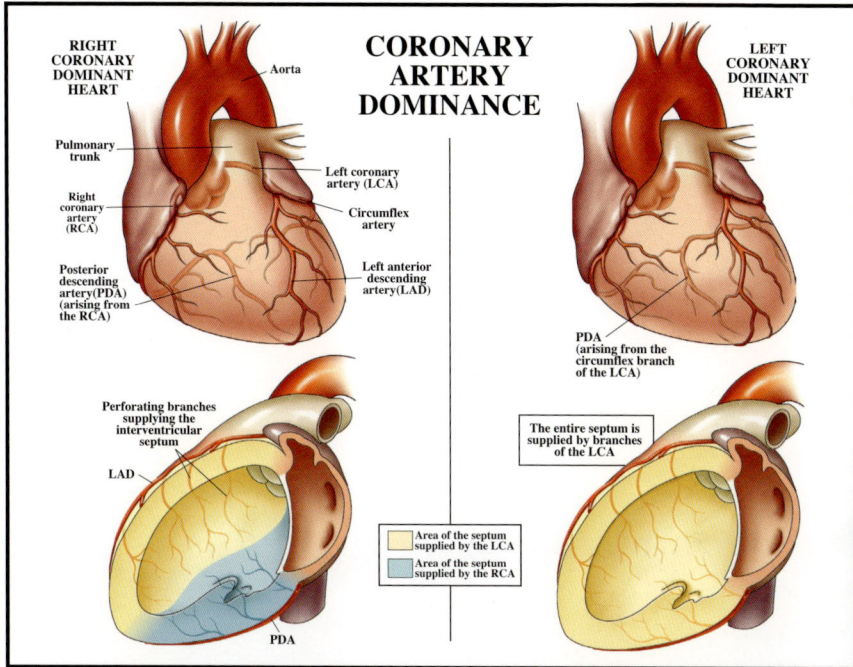

CORONARY ARTERY DOMINANCE

RIGHT CORONARY DOMINANT HEART

Aorta

Pulmonary trunk

Left coronary artery (LCA)

Right coronary artery (RCA)

Circumflex artery

Posterior descending artery(PDA) (arising from the RCA)

Left anterior descending artery(LAD)

LEFT CORONARY DOMINANT HEART

PDA (arising from the circumflex branch of the LCA)

Perforating branches supplying the interventricular septum

LAD

The entire septum is supplied by branches of the LCA

Area of the septum supplied by the LCA

Area of the septum supplied by the RCA

PDA

C3: Coronary artery dominance

• Right heart dominance: the posterior portion of the interventricular septum is supplied by the posterior descending branch of the right coronary artery.

• Left heart dominance: the entire septum is supplied by branches of the left anterior descending artery; an obstruction in that vessel may lead to loss of the entire septum, an often fatal event. The posterior descending artery is derived from a branch of the circumflex artery instead of from the RCA.

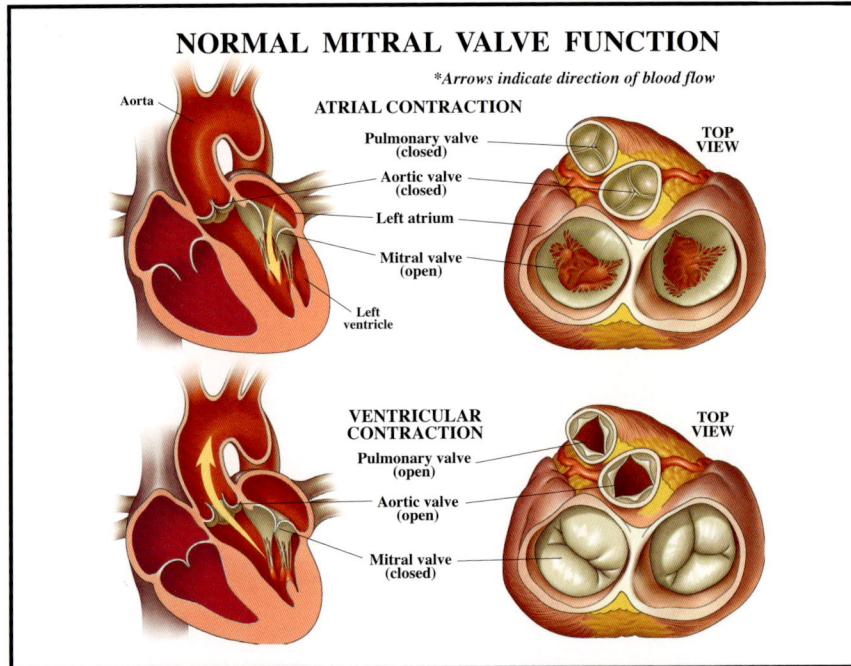

NORMAL MITRAL VALVE FUNCTION

Arrows indicate direction of blood flow

ATRIAL CONTRACTION

Aorta
Pulmonary valve (closed)
Aortic valve (closed)
Left atrium
Mitral valve (open)
Left ventricle

TOP VIEW

VENTRICULAR CONTRACTION

Pulmonary valve (open)
Aortic valve (open)
Mitral valve (closed)

TOP VIEW

C4: Normal mitral valve function

• Also known as the left atrioventricular valve, the mitral valve has 2 leaflets which are anchored to the ventricle floor by papillary muscles and chordae tendinae, as are the leaflets of the right atrioventricular valve (tricuspid valve).

• The aortic and pulmonary (pulmonic) valves are semilunar valves, and have thin cusps with thickened edges which seal during diastole (when the ventricles are relaxed and blood flows into the atria).

• The mitral valve prevents backflow from the left ventricle into the atrium; minor mitral valve prolapse or leak is usually clinically insignificant.

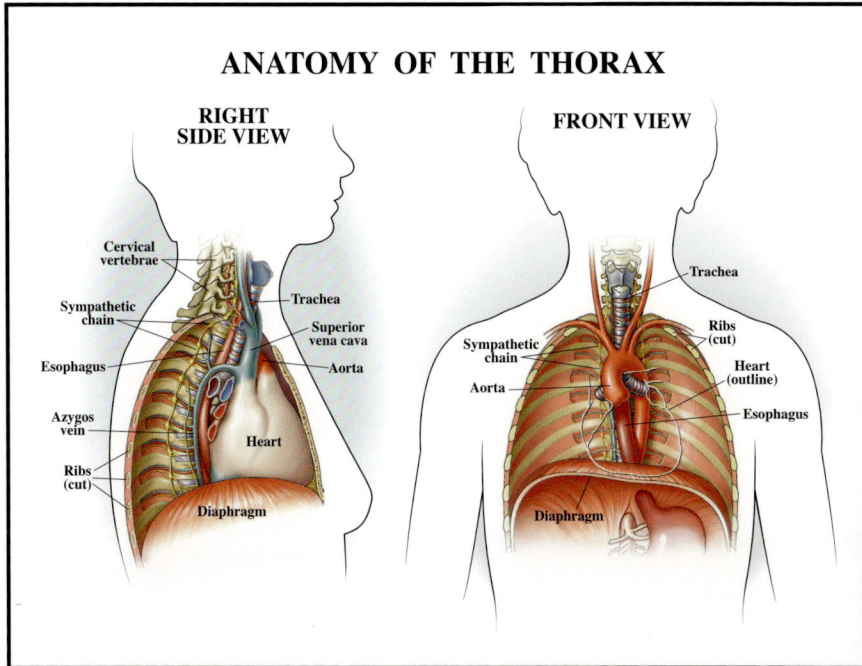

ANATOMY OF THE THORAX

RIGHT SIDE VIEW

Cervical vertebrae
Sympathetic chain
Esophagus
Azygos vein
Ribs (cut)
Trachea
Superior vena cava
Aorta
Heart
Diaphragm

FRONT VIEW

Trachea
Ribs (cut)
Heart (outline)
Esophagus
Sympathetic chain
Aorta
Diaphragm

C5: Anatomy of the thorax

- The thoracic structures are in a relatively closed-in, dome-shaped space, with the ribs, costal cartilage and vertebral column constituting the bony sidewalls; the base is formed by the diaphragm.

- The lungs within their pleural sacs fill each side of the thorax. The mediastinum contains the heart, trachea, esophagus, aortic arch and descending aorta, the terminal portions of the great veins and the thymus gland.

- The heart is surrounded by the sturdy pericardial sac, made of a nondistensible fibrous material.

- The esophagus lies against the vertebral column posteriorly.

- The great vessels and esophagus pass through openings in the diaphragm into the abdomen.

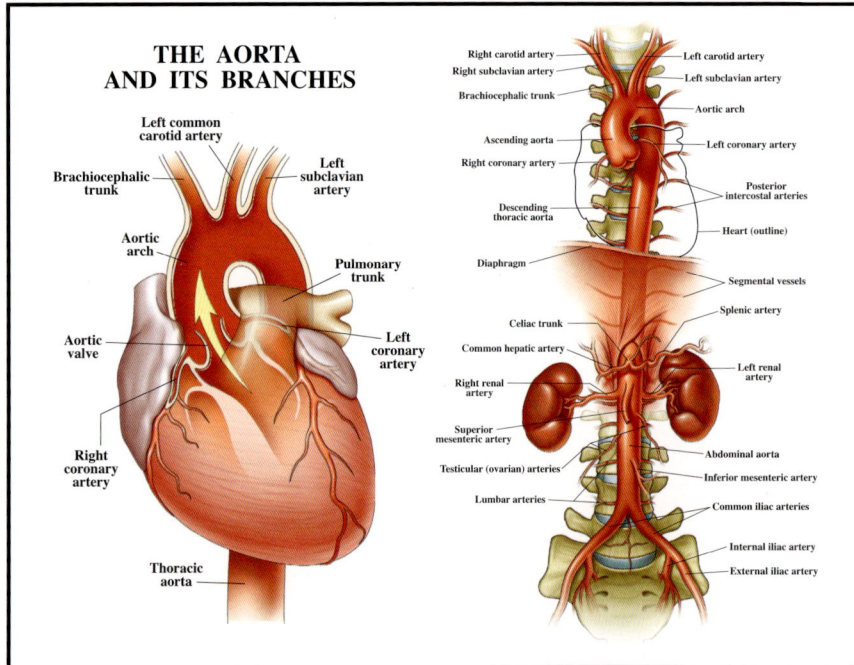

THE AORTA AND ITS BRANCHES

Left common carotid artery
Brachiocephalic trunk
Left subclavian artery
Aortic arch
Pulmonary trunk
Aortic valve
Left coronary artery
Right coronary artery
Thoracic aorta

Right carotid artery
Right subclavian artery
Brachiocephalic trunk
Ascending aorta
Right coronary artery
Descending thoracic aorta
Diaphragm
Celiac trunk
Common hepatic artery
Right renal artery
Superior mesenteric artery
Testicular (ovarian) arteries
Lumbar arteries

Left carotid artery
Left subclavian artery
Aortic arch
Left coronary artery
Posterior intercostal arteries
Heart (outline)
Segmental vessels
Splenic artery
Left renal artery
Abdominal aorta
Inferior mesenteric artery
Common iliac arteries
Internal iliac artery
External iliac artery

C6: Aorta and its branches

- The largest artery in the body, the aorta leaves the left ventricle and supplies all of the body's tissues, including the heart itself, via its branches.

- The major parts are the arch (from the aortic valve to the left subclavian artery), descending thoracic aorta (from the left subclavian to the diaphragm) and abdominal aorta (from the diaphragm to the iliac bifurcation at about the level of the umbilicus).

- There are three major vessels leaving the arch and these supply the head, neck and arms; there are segmental vessels supplying the body wall throughout the length of the aorta. Major branches supply abdominal structures; the iliac arteries and their branches supply the pelvis and legs.

ELECTROCARDIOGRAPHY

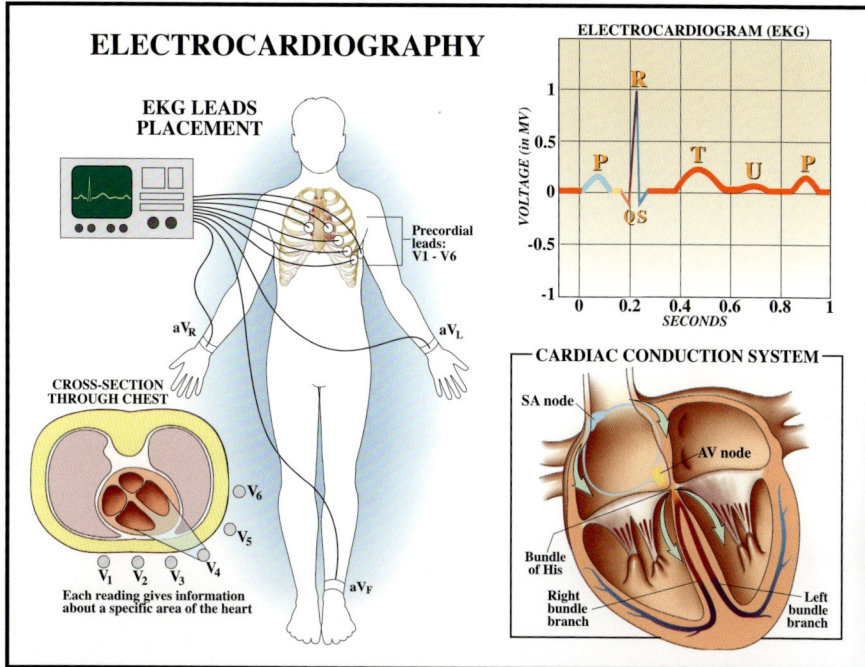

EKG LEADS PLACEMENT

Precordial leads: V1 - V6

aV$_R$

aV$_L$

aV$_F$

V$_6$
V$_5$
V$_4$
V$_3$
V$_2$
V$_1$

CROSS-SECTION THROUGH CHEST

Each reading gives information about a specific area of the heart

ELECTROCARDIOGRAM (EKG)

VOLTAGE (in MV)

R

P   T   U   P

QS

SECONDS

CARDIAC CONDUCTION SYSTEM

SA node

AV node

Bundle of His

Right bundle branch

Left bundle branch

C7: Electrocardiography

• A tracing is made from electrical impulses traveling through the heart, tracking the way the heart muscle reacts to the conduction system.

• An electrical impulse is initiated at the sinoatrial node, passes through specialized neuromuscular fibers lying beneath the inner lining of the heart until it reaches the atrioventricular node; from there, it travels through the Bundle of His, into the bundle branches and the Purkinje fibers, stimulating ventricular contraction.

• Changes in tracings are evaluated by comparing them to normal and/or baseline tracings; a physician can get information about areas of heart damage, both acute and chronic.

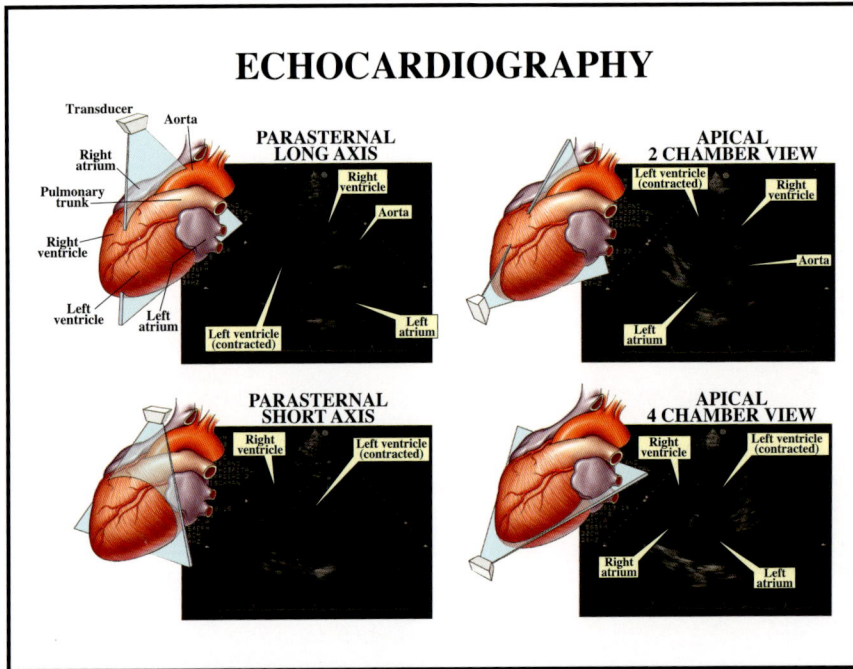

ECHOCARDIOGRAPHY

C8: Echocardiography

- Using sound waves, this technique takes a "picture" of slices of the heart as the heart beats.

- The equipment can be calibrated to show muscle movement and blood flow in real time.

- Echocardiography can be done from outside of the body wall or through a probe placed within the esophagus (TEE or transesophageal echocardiogram).

- The tests can determine anatomy, blood flow, and whether all parts of the heart are functioning properly, in real time.

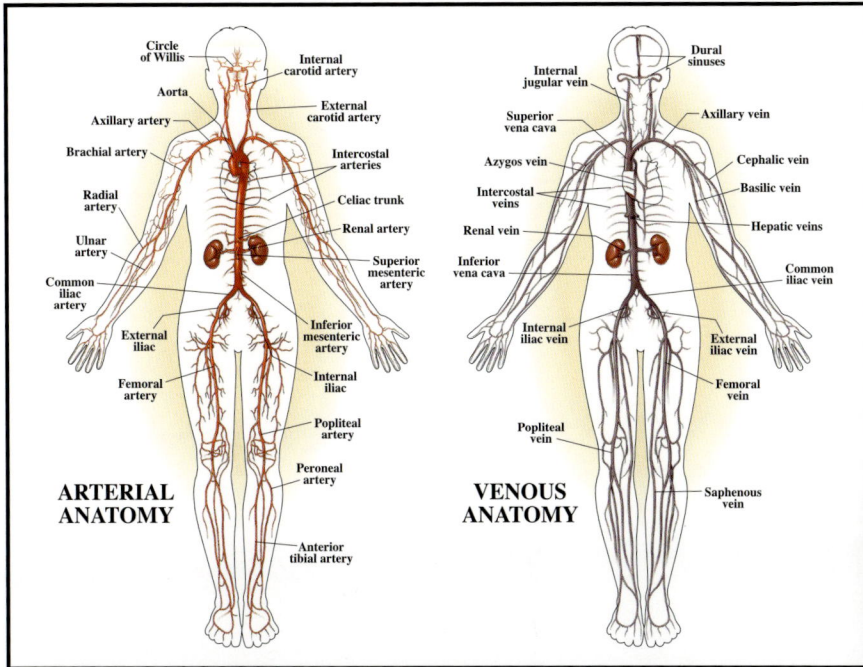

ARTERIAL ANATOMY

Circle of Willis · Internal carotid artery · Aorta · External carotid artery · Axillary artery · Brachial artery · Intercostal arteries · Radial artery · Celiac trunk · Renal artery · Ulnar artery · Superior mesenteric artery · Common iliac artery · External iliac · Inferior mesenteric artery · Femoral artery · Internal iliac · Popliteal artery · Peroneal artery · Anterior tibial artery

VENOUS ANATOMY

Internal jugular vein · Dural sinuses · Superior vena cava · Axillary vein · Azygos vein · Cephalic vein · Intercostal veins · Basilic vein · Renal vein · Hepatic veins · Inferior vena cava · Common iliac vein · Internal iliac vein · External iliac vein · Femoral vein · Popliteal vein · Saphenous vein

C9: Arterial and venous systems

• Although the cardiovascular system is referred to as one unit, it is actually two separate systems which work independently.

• Through the arterial supply, oxygenated blood is distributed from the lungs to the left heart and aorta, and eventually to within 5 cells of every cell in the body. The arteries divide into smaller arteries, then into arterioles, which in turn divide into capillaries. Oxygen exchange takes place at the level of the capillaries, vessels whose walls are only one cell thick.

• In the venous system, deoxygenated blood drains from the capillaries, which conjoin into venules, small veins, veins, and the major draining vessels - the superior and inferior venae cavae. This blood then enters the right heart and travels to the lungs to re-oxygenate and start the cycle again.

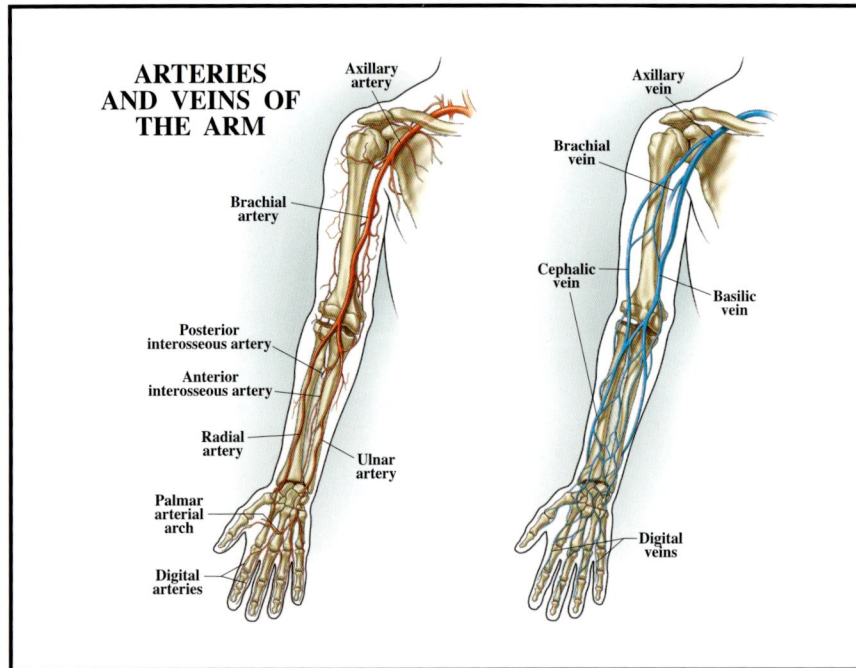

ARTERIES AND VEINS OF THE ARM

Axillary artery

Brachial artery

Posterior interosseous artery

Anterior interosseous artery

Radial artery

Palmar arterial arch

Digital arteries

Ulnar artery

Axillary vein

Brachial vein

Cephalic vein

Basilic vein

Digital veins

C10: Arteries/veins of the arm

- Most arteries and veins in the body travel and are named together; the exceptions include the superficial veins of the extremities.

- The arterial supply to the arm comes from one vessel branching off the aorta (subclavian artery on the left, and brachiocephalic trunk on the right). The name of the artery changes as it travels past certain landmarks; it sends smaller branches along its length until dividing into the two major branches near the elbow.

- The terminal branches of the ulnar and radial arteries communicate with each other via two arches. These assure collateral circulation to the hands and fingers in case of damage.

- The veins seen through the skin of the arm are unpaired superficial veins and have no analogous arteries.

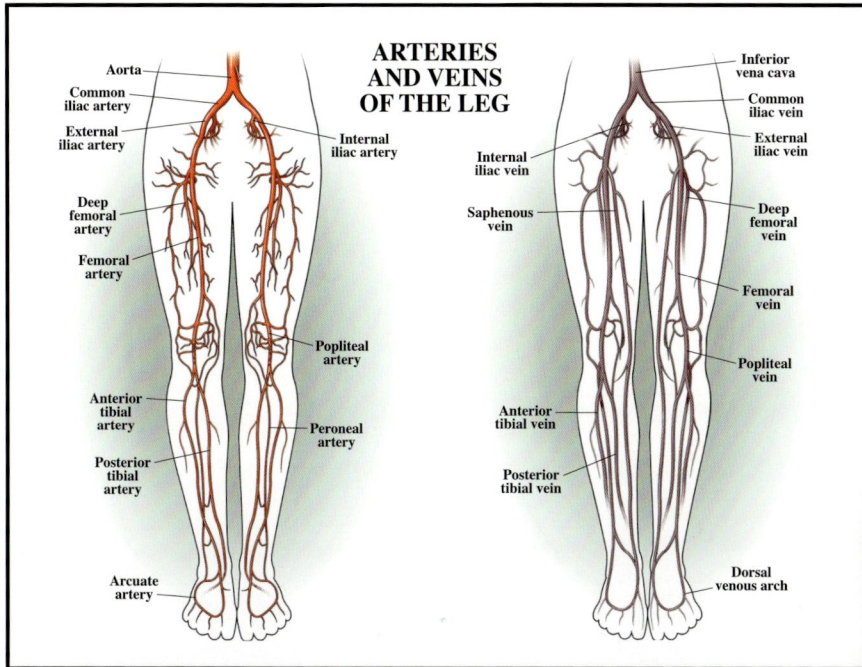

ARTERIES AND VEINS OF THE LEG

Aorta
Common iliac artery
External iliac artery
Internal iliac artery
Deep femoral artery
Femoral artery
Popliteal artery
Anterior tibial artery
Peroneal artery
Posterior tibial artery
Arcuate artery

Inferior vena cava
Common iliac vein
External iliac vein
Internal iliac vein
Saphenous vein
Deep femoral vein
Femoral vein
Popliteal vein
Anterior tibial vein
Posterior tibial vein
Dorsal venous arch

C11: Arteries/veins of the leg

- Like in the arm, the deep, muscular arteries of the leg travel together, but the superficial veins are unpaired and variable in course.

- The legs receive blood from the terminal branches of the aorta, the iliac arteries. Branches supply the muscles of the thigh, and the name of the artery changes as it passes certain landmarks.

- The major vessels trifurcate behind the knee, dividing into the anterior and posterior tibial arteries and the peroneal artery, all of which travel toward the feet supplying the muscles and other tissues along the way.

- As in the arm, there are connective vascular arches in the foot supplying collateral circulation.

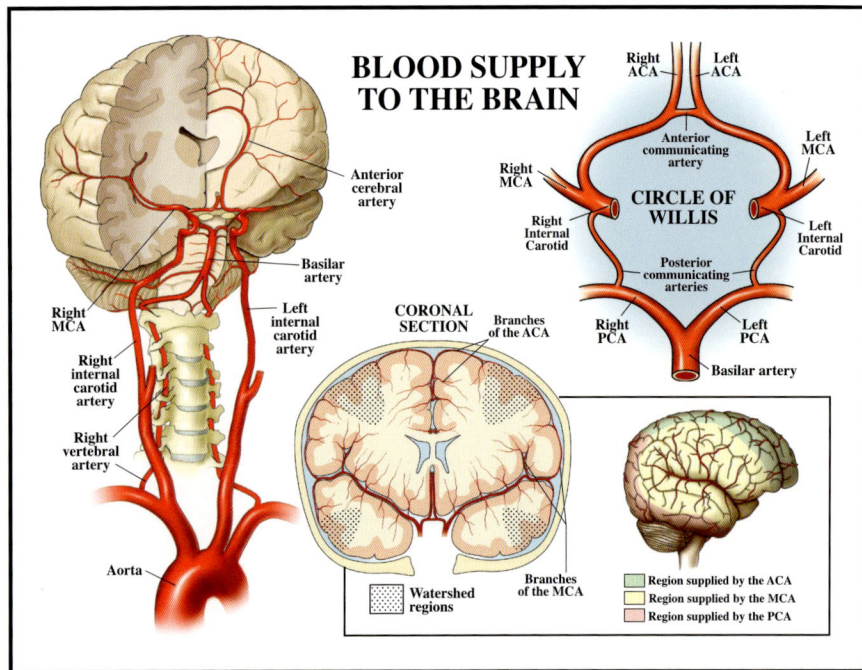

BLOOD SUPPLY
TO THE BRAIN

Anterior
cerebral
artery

Basilar
artery

Right
MCA

Left
internal
carotid
artery

Right
internal
carotid
artery

Right
vertebral
artery

Aorta

Right
ACA

Left
ACA

Anterior
communicating
artery

Right
MCA

Left
MCA

CIRCLE OF
WILLIS

Right
Internal
Carotid

Left
Internal
Carotid

Posterior
communicating
arteries

Right
PCA

Left
PCA

Basilar artery

CORONAL
SECTION

Branches
of the ACA

Branches
of the MCA

Watershed
regions

Region supplied by the ACA
Region supplied by the MCA
Region supplied by the PCA

C12: Blood supply to brain

- The anterior 2/3 of the brain is supplied by branches of the internal carotid artery, whose terminal branches form the anterior and middle cerebral arteries.

- The vertebral arteries branch off the subclavian arteries and through small openings in the transverse processes of the cervical vertebrae. They merge to form the basilar artery supplying the cerebellum, brain stem and the posterior cerebrum via the posterior cerebral arteries.

- The Circle of Willis has small connecting vessels between the three major cerebral vessels. Blood can change direction within the circle for collateral blood flow if needed. There can be significant variation in the form of the Circle of Willis from one individual to another.

- Tissues supplied by the tiny terminal branches of vessels are known as watershed regions, and are vulnerable to damage during periods of low perfusion or oxygenation.

AORTIC DISSECTION

ANTERIOR VIEW

SECTIONED VIEW

A.

Aortic arch

Adventitia

Media

Intima

Intimal tear

Lumen

Aortic valve

B.

Blood forces its way up into the tear, causing the intima to separate

A false lumen is created

Enlarged aortic diameter

True lumen narrowed

Site of intimal tear

False lumen

True lumen

In some areas, the false lumen can obliterate the true lumen, compromising blood flow

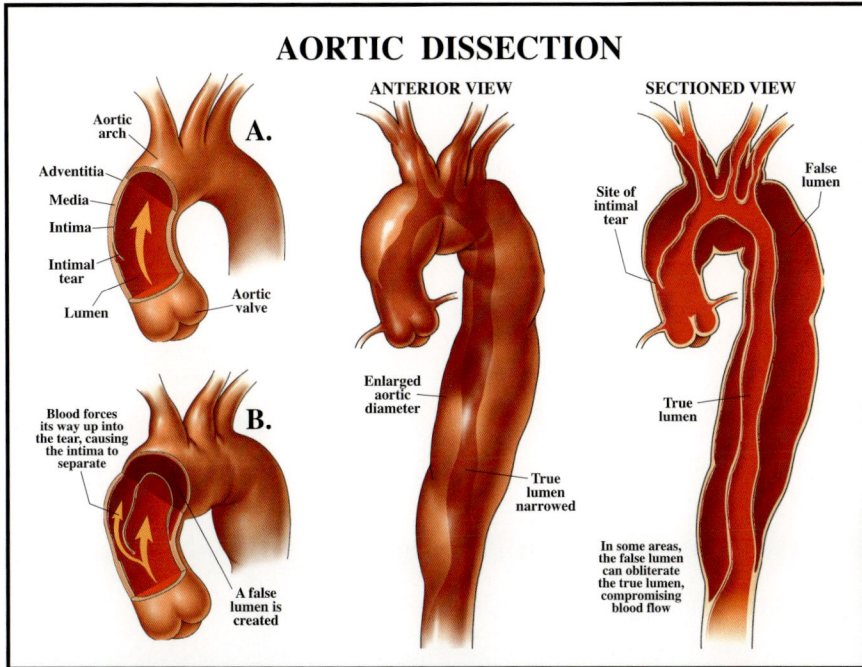

C13: Aortic dissection

- Sometimes called "dissecting aneurysm", this is not an aneurysm, but a separation of the aortic wall layers.

- Blood enters the aortic wall through a small tear in the intima, or inner lining of the artery. Under pressure, it then dissects through the wall, creating a false lumen or false channel. Sometimes there is a second tear through which the blood re-enters the true aortic lumen; some-times the blood breaks through the wall to the thorax or retroperitoneal spaces.

- Dissections are usually associated with hypertension and atherosclerosis, although certain genetic conditions (Marfan's syndrome) can predispose to dissection.

- Symptoms include a severe tearing pain in the back as the dissection travels distally, changes in blood pressure and distal pulses, and loss of various physi-ologic functions if the dissection blocks the blood supply to major organs.

13

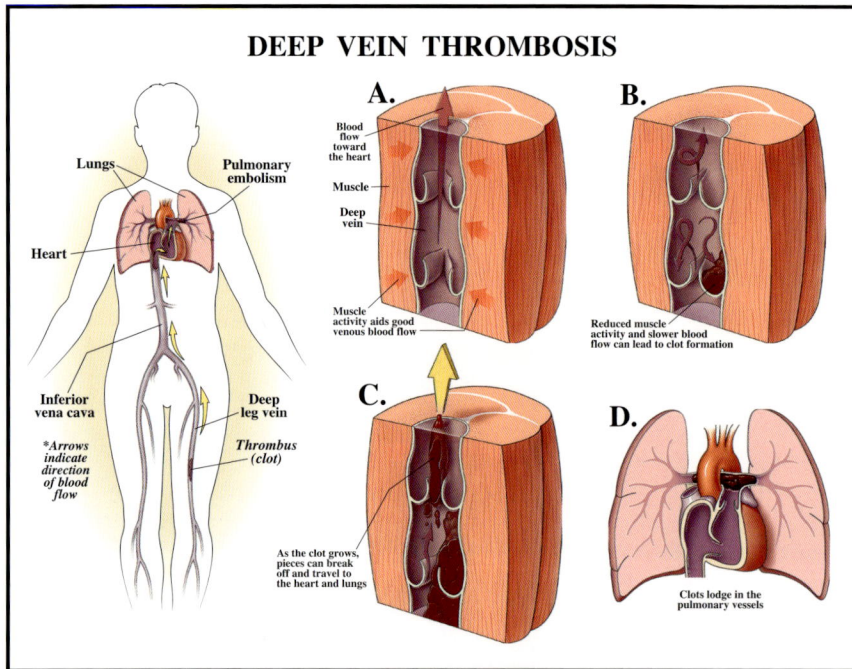

DEEP VEIN THROMBOSIS

Lungs

Pulmonary embolism

Heart

Inferior vena cava

Deep leg vein

Arrows indicate direction of blood flow

Thrombus (clot)

A.

Blood flow toward the heart

Muscle

Deep vein

Muscle activity aids good venous blood flow

B.

Reduced muscle activity and slower blood flow can lead to clot formation

C.

As the clot grows, pieces can break off and travel to the heart and lungs

D.

Clots lodge in the pulmonary vessels

C14: Deep vein thrombosis

- These potentially lethal blood clots usually form in the deep veins of the leg or in the pelvis. Those in the legs are usually painful, whereas those in ths pelvis may be asymptomatic.

- The clots usually form in the valves of the larger veins, propagating upward toward the heart.

- Clots of any size can break off and travel with the blood flow through the inferior vena cava to the right side of the heart and to the lungs; the pulmonary vasculature acts like a sieve and clots get caught in the vessels as the vessels get smaller, causing loss of blood flow in those areas. These clots are known as pulmonary emboli if they reach the lungs.

- Conditions associated with DVT and PE include history of leg trauma, cancer, surgery, venous stasis from illness, lack of exercise, clotting defects and others.

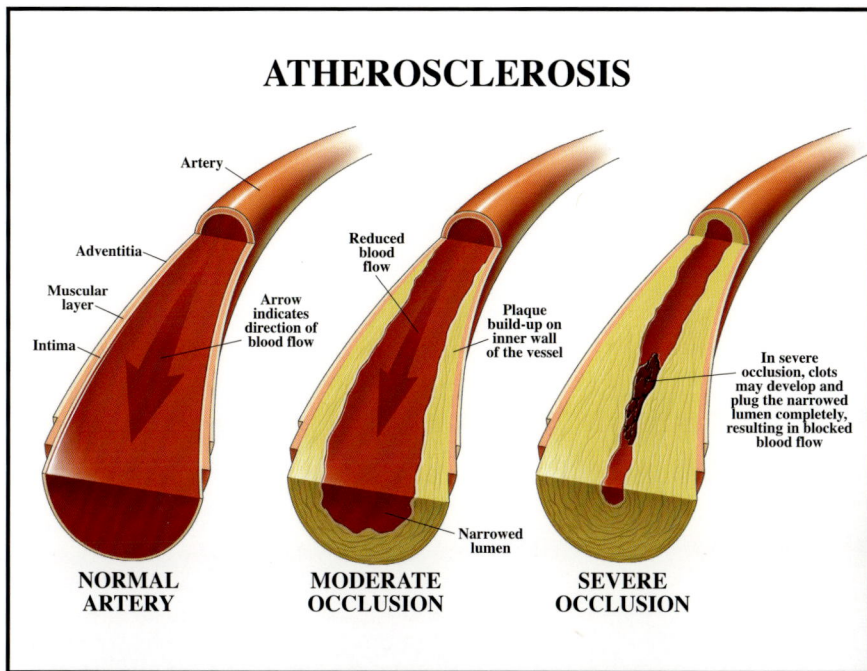

ATHEROSCLEROSIS

Artery

Adventitia

Muscular layer

Intima

Arrow indicates direction of blood flow

Reduced blood flow

Plaque build-up on inner wall of the vessel

Narrowed lumen

In severe occlusion, clots may develop and plug the narrowed lumen completely, resulting in blocked blood flow

NORMAL ARTERY

MODERATE OCCLUSION

SEVERE OCCLUSION

C15: Atherosclerosis

• Atherosclerotic plaque is fatty, cholesterol-laden material which accumulates within the inner layer of the major arteries, narrowing the diameter of the lumen or opening.

• It can occur in any artery in the body and is a direct cause of stroke when in the carotid arteries; myocardial infarction when in the coronary arteries; acute bowel ischemia when in the mesenteric vessels; peripheral vascular disease when in vessels to the legs, etc.

• Atherosclerosis can result in increased blood pressure in an effort to overcome the higher pressures caused by arterial stenosis throughout the body.

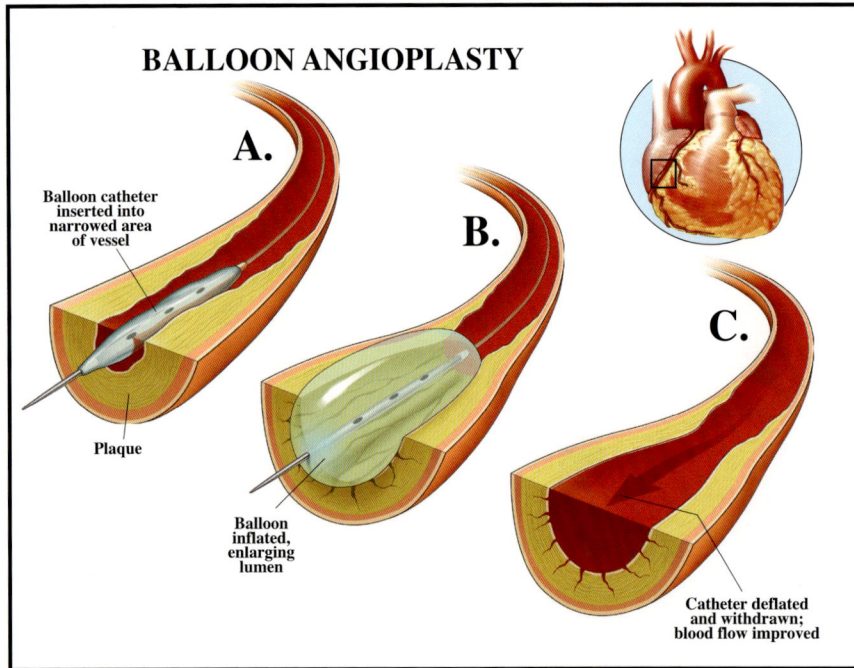

BALLOON ANGIOPLASTY

A.

Balloon catheter inserted into narrowed area of vessel

Plaque

B.

Balloon inflated, enlarging lumen

C.

Catheter deflated and withdrawn; blood flow improved

C16: Balloon angioplasty

- This procedure is a relatively non-invasive technique of opening stenotic blood vessels.

- A catheter is threaded through the arterial system from the arm or leg and into the diseased artery. The balloon is then positioned inside the stenotic area and gently inflated several times to crush the plaque and flatten it against the walls of the vessel.

- This procedure is commonly performed and is often accompanied by deployment of a stent to hold the vessel open.

- Complications can include clot formation on the fractured plaque after release of clotting factors and formation of a dissection (sometimes incorrectly called a "dissecting aneurysm") in the vessel wall.

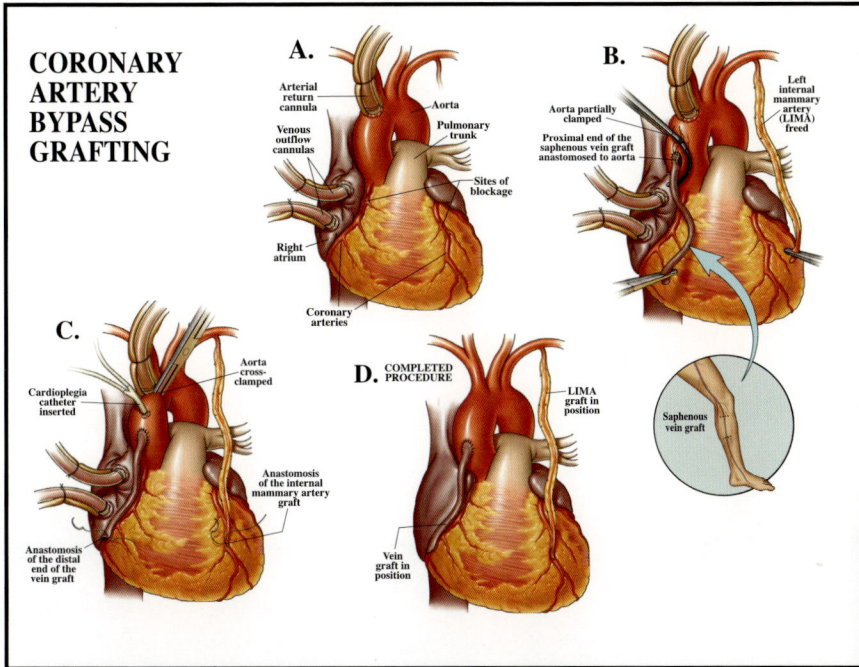

CORONARY
ARTERY
BYPASS
GRAFTING

A.
Arterial return cannula
Aorta
Venous outflow cannulas
Pulmonary trunk
Sites of blockage
Right atrium
Coronary arteries

B.
Aorta partially clamped
Proximal end of the saphenous vein graft anastomosed to aorta
Left internal mammary artery (LIMA) freed

C.
Cardioplegia catheter inserted
Aorta cross-clamped
Anastomosis of the internal mammary artery graft
Anastomosis of the distal end of the vein graft

D. COMPLETED PROCEDURE
LIMA graft in position
Vein graft in position

Saphenous vein graft

C17: Coronary artery bypass grafting

- When coronary arteries are significantly blocked (>70% stenosis), pain symptoms often occur with exercise in the form of stable angina, or at rest in the form of unstable angina.

- Bypass vessels are harvested as free grafts from the saphenous veins of the legs. These are anastomosed to the aorta and then to the coronary arteries, literally bypassing the blocked regions.

- The internal mammary arteries, which lie on either side of the sternum within the ribcage, can also be harvested and anastomosed directly to the coronary arteries.These grafts are less likely to stenose than are vein grafts.

- The procedure can be performed either on or off cardiopulmonary bypass.

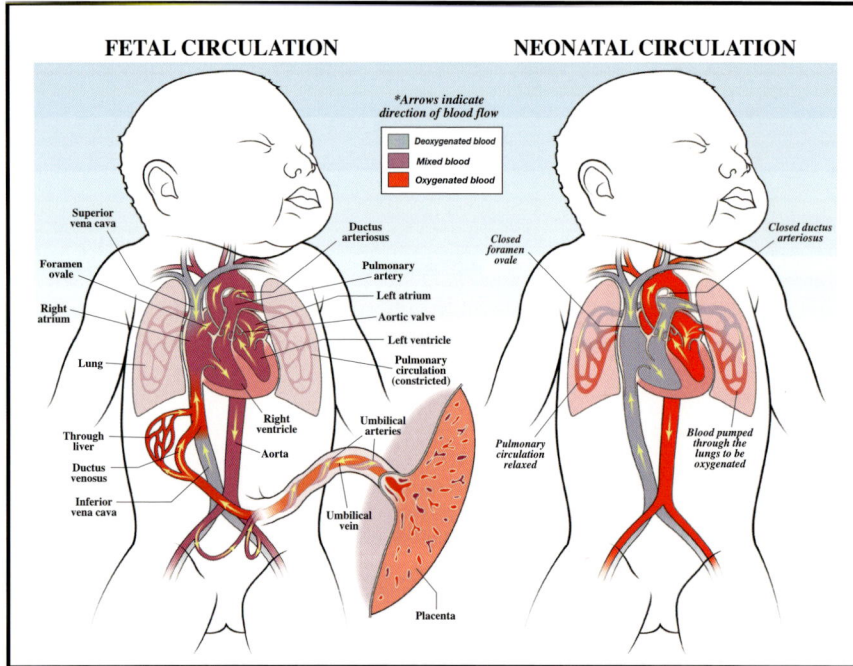

FETAL CIRCULATION

NEONATAL CIRCULATION

*Arrows indicate direction of blood flow

Deoxygenated blood
Mixed blood
Oxygenated blood

Superior vena cava
Foramen ovale
Right atrium
Lung
Through liver
Ductus venosus
Inferior vena cava

Ductus arteriosus
Pulmonary artery
Left atrium
Aortic valve
Left ventricle
Pulmonary circulation (constricted)
Right ventricle
Umbilical arteries
Aorta
Umbilical vein
Placenta

Closed foramen ovale
Pulmonary circulation relaxed

Closed ductus arteriosus
Blood pumped through the lungs to be oxygenated

C18: Fetal/neonatal circulation

• The growing fetus receives its oxgen from the mother so has no need for functioning lungs. The oxygenated blood enters the fetus through the umbilical vein and travels through the liver to the heart.

• Once the blood reaches the right heart, it is shunted directly into the left atrium for distribution through the aorta. There is a physiological opening between the atria, (foramen ovale) and a second vascular connection between the pulmonary artery and the aorta (ductus arteriosus), through which blood shunts.

• Soon after delivery, the lungs expand, the foramen ovale and ductus arteriosus close, and blood begins to circulate through the lungs to supply oxygen for the tissues. Failure to convert to neonatal circulation is known as persistent fetal circulation and must be treated.

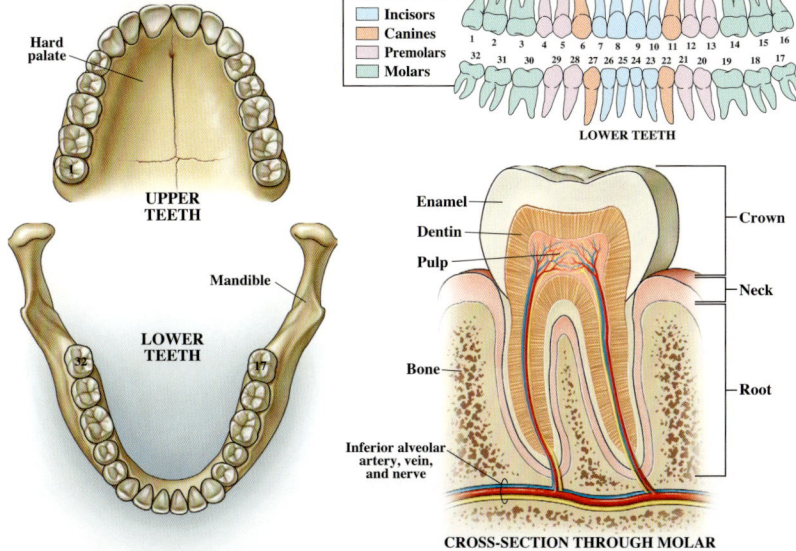

NORMAL DENTITION

Hard palate

UPPER TEETH

Mandible

LOWER TEETH

UPPER TEETH

- Incisors
- Canines
- Premolars
- Molars

1 2 3 4 5 6 7 8 9 10 11 12 13 14 15 16
32 31 30 29 28 27 26 25 24 23 22 21 20 19 18 17

LOWER TEETH

Enamel
Dentin
Pulp

Bone

Inferior alveolar artery, vein, and nerve

Crown

Neck

Root

CROSS-SECTION THROUGH MOLAR

D1: Normal dentition

- There are normally 32 teeth, divided into 4 categories: molars, premolars, canines and incisors.

- The teeth are embedded in the bones of the maxilla (upper jaw) and mandible (lower jaw) and are held in position by periodontal ligaments.

- The third molars (teeth #1, 16, 17, 32) are often vestigial and/or impacted. These are commonly known as "wisdom teeth".

- The roots of the teeth are anchored within the bone and contain an artery, vein and nerve which travel to the main portion of the tooth and divide within the pulp.

- Dentin covers the pulp and very hard enamel covers the dentin; the bone is covered with a mucosal tissue, the gingiva.

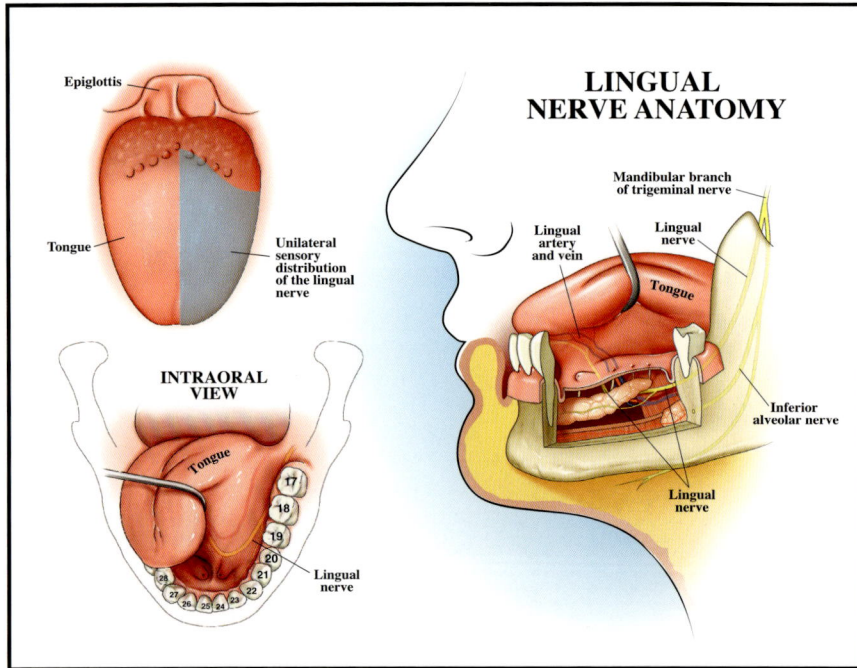

Epiglottis

Tongue

Unilateral sensory distribution of the lingual nerve

INTRAORAL VIEW

Tongue

17
18
19
20
21
22
23 24 25 26 27 28

Lingual nerve

LINGUAL NERVE ANATOMY

Mandibular branch of trigeminal nerve

Lingual artery and vein

Lingual nerve

Tongue

Inferior alveolar nerve

Lingual nerve

D2: Lingual nerve anatomy

• The lingual nerve supplies sensation to the anterior 2/3 of the tongue.

• A branch of the mandibular nerve (the third portion of the trigeminal nerve), it usually travels along the inner surface of the mandible, under the gingiva, then turns medially to supply the tongue.

• The lingual nerve is occasionally damaged during removal of the third molars ("wisdom teeth") if it has a variation in course and travels over the alveolar ridge. When the gingiva is incised to access the tooth, the nerve can be damaged or transected. There is no way to determine the location of the nerve prior to surgery.

PERIODONTAL DISEASE

*Periodontal disease is caused
by bacteria and is fostered by poor oral hygiene*

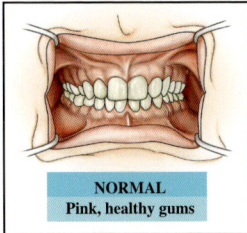

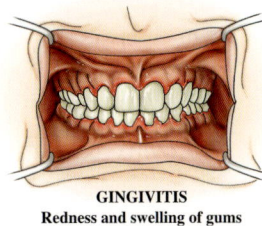

NORMAL
Pink, healthy gums

GINGIVITIS
Redness and swelling of gums

II. GINGIVITIS:
- Chronic inflammation of gingival margin
- Bleeding gums
- Discomfort upon eating or brushing

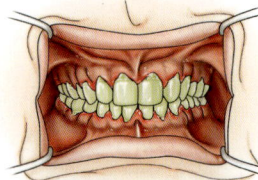

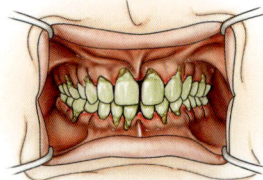

II. PERIODONTITIS:
- Deepened gingival crevice or periodontal pocket
- Pus formation within periodontal pocket
- Recession of gums to expose tooth roots
- Tooth migration
- Alveolar bone loss
- Tooth loss

EARLY PERIODONTITIS
Deepening of periodontal pocket

LATE PERIODONTITIS
Pus formation, exposure of tooth root

D3: Periodontal disease

- One of the major causes of tooth loss is periodontal disease. Starting as an inflammation of the gums (gingivitis), it progresses to undermining of the gingiva, and deepening of soft-tissue "pockets" surrounding the roots. Periodontal disease progresses from gum inflammation, to gum recession and then root exposure and loss of periodontal ligaments which normally support the teeth within the alveolar bone.

- Gum and tooth loss are accompanied by alveolar bone recession.

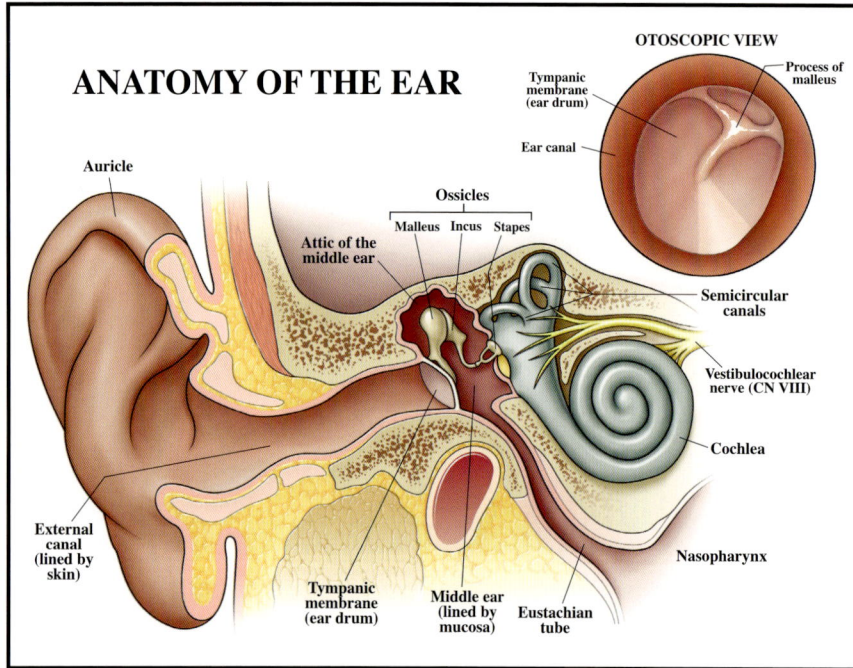

ANATOMY OF THE EAR

OTOSCOPIC VIEW

Tympanic membrane (ear drum)

Process of malleus

Ear canal

Auricle

Ossicles

Malleus Incus Stapes

Attic of the middle ear

Semicircular canals

Vestibulocochlear nerve (CN VIII)

Cochlea

External canal (lined by skin)

Nasopharynx

Tympanic membrane (ear drum)

Middle ear (lined by mucosa)

Eustachian tube

E1: Anatomy of the ear

• The external ear acts as a collecting device for sound waves, focusing them into the canal.

• External sound waves cause vibration of the tympanic membrane (ear drum). The vibrating membrane moves the three ossicles of the middle ear (malleus, incus and stapes) which transfer the vibration to branches of the vestibulocochlear nerve (cranial nerve VIII) within the cochlea.

• Motion and balance are detected by three fluid-filled canals in the temporal bone. Oriented in three perpendicular planes, the canals contain tiny hair cells that pick up fluid movement with motion of the head. This information is transmitted through the vestibular portions of the nerve to the appropriate portions of the brain.

ABDOMINAL ANATOMY

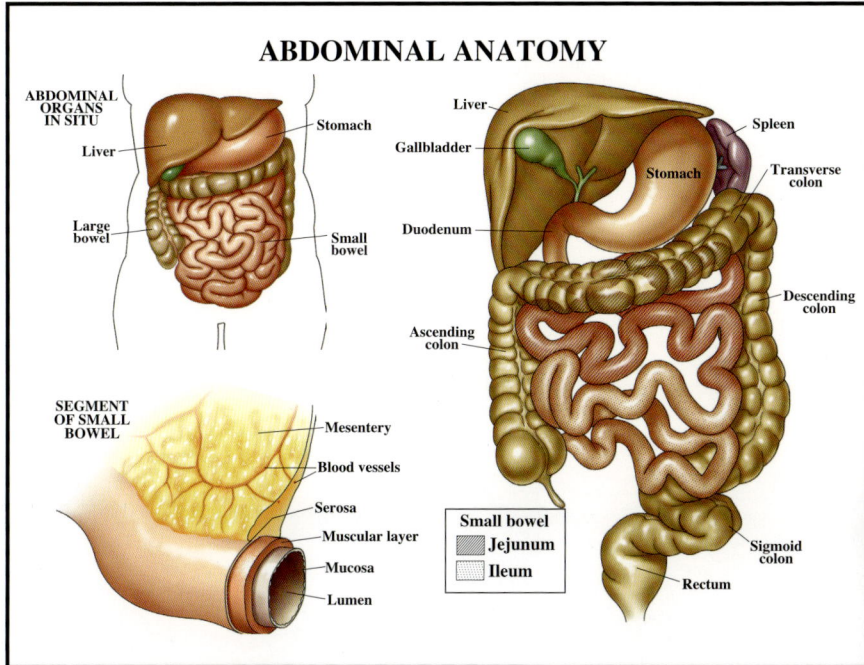

ABDOMINAL ORGANS IN SITU

- Liver
- Stomach
- Large bowel
- Small bowel

SEGMENT OF SMALL BOWEL

- Mesentery
- Blood vessels
- Serosa
- Muscular layer
- Mucosa
- Lumen

- Liver
- Gallbladder
- Stomach
- Spleen
- Transverse colon
- Duodenum
- Descending colon
- Ascending colon
- Sigmoid colon
- Rectum

Small bowel
- Jejunum
- Ileum

GI1: Abdominal anatomy

- The contents of the abdomen are primarily associated with digestion and distribution of nutrients.

- The esophagus, a tube which carries food and fluid through the thorax, enters the abdomen through the diaphragm, where it widens into the stomach; the stomach empties into the small bowel (duodenum, jejunum and ileum, in which food is absorbed into the blood stream), and from there into the large bowel, where waste material is compacted as fluid is reabsorbed into the system.

- The liver has multiple functions affecting a number of other body systems, including digestive, hematologic and endocrine/metabolic.

- The large and small bowels are supplied by branches off of the aorta carried within the mesentery, a double-layered sheetlike structure.

23

UPPER ABDOMINAL ANATOMY

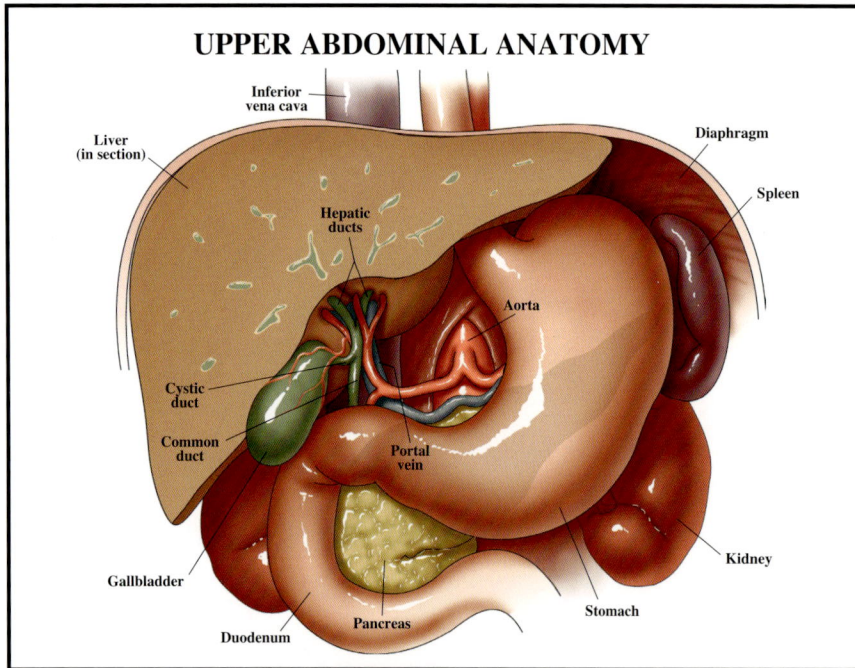

Inferior vena cava

Liver (in section)

Diaphragm

Spleen

Hepatic ducts

Aorta

Cystic duct

Common duct

Portal vein

Gallbladder

Kidney

Duodenum

Pancreas

Stomach

GI2: Upper abdominal anatomy

- The primary function of the upper abdominal organs is the breakdown of food for distribution by the small bowel. Chewed and macerated food travels through the esophagus to the stomach, where strong acids and muscular contractions break it down further.

- Proteolytic enzymes from the pancreas and bile from the liver and gallbladder drain into the duodenum to further the digestion and breakdown of food.

- The spleen functions as part of the hematopoietic system, controlling the distribution and eventual destruction of red blood cells. It also acts as a part of the immune system.

- Blood is supplied to most of these structures by branches of the celiac trunk, the first major aortic branch in the abdomen.

STOMACH ANATOMY

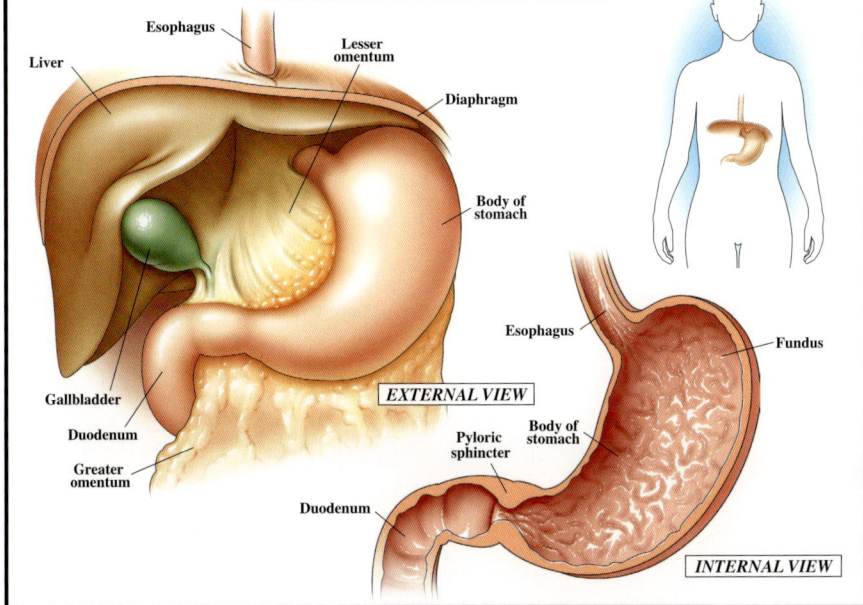

Esophagus

Liver

Lesser omentum

Diaphragm

Body of stomach

Gallbladder

Duodenum

Greater omentum

EXTERNAL VIEW

Esophagus

Fundus

Pyloric sphincter

Body of stomach

Duodenum

INTERNAL VIEW

GI3: Stomach anatomy

• The stomach is a muscular sac derived from the simple fetal gastrointestinal tube. The mucosal lining has specialized cells which secrete strong acids and enzymes to break food down before it passes to the small bowel for absorption and distribution.

• The walls are folded into rugae which increase the surface area of the sac. The muscular walls contract to help break up food material.

• The greater omentum arises from the greater curvature of the stomach, and the lesser omentum from the lesser curvature; the hepatoduodenal ligament lies at the free edge and contains the extrahepatic biliary ducts.

• The stomach lies under the diaphragm and to the left of the liver. The strong pyloric sphincter divides the distal stomach from the duodenum, or the first portion of the small intestine.

BILIARY PHYSIOLOGY

Intrahepatic ducts

(1) *Liver secretes bile into bile ducts*

Liver

Left and right hepatic ducts

Gallbladder

(2) *Bile is stored in gallbladder where it is concentrated*

Gallbladder

Common duct

Cystic duct

(3) *Upon ingestion of fat, the duodenum releases cholecystokinin, which causes the gallbladder to contract and the sphincter of Oddi to relax*

Cystic duct

(4) *Bile then flows from the gallbladder into the duodenum to break down fat*

Duodenum

Common duct

Ingested fat

Duodenum

Sphincter of Oddi

Pancreas

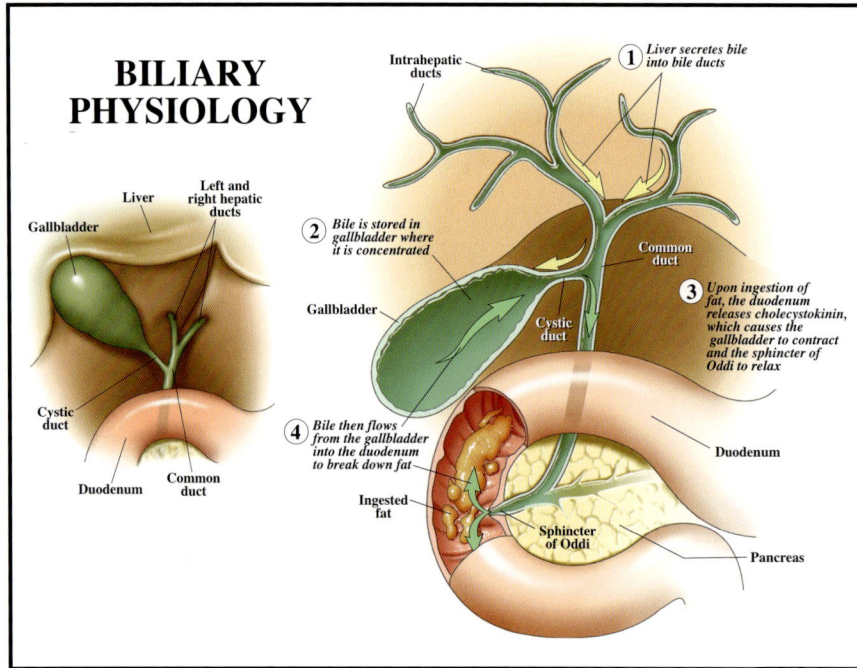

GI4: Biliary physiology

• The gallbladder stores bile formed within the liver, releasing it for fat digestion.

• Bile travels through the intrahepatic ducts into the paired hepatic ducts; these merge into the common hepatic duct. Bile is then diverted via the cystic duct to the gallbladder for storage.

• When food is ingested and travels through the stomach to the duodenum, a hormone is released (cholecystokinin) which stimulates the gallbladder to contract and the sphincter of Oddi to relax. This allows bile to flow through the cystic duct and the common bile duct into the duodenum.

• The most common pathology in the extrahepatic biliary system is bile (gall) stones (concretions of bile salts, cholesterol and minerals) which can block ducts, causing inflammation, pain and jaundice.

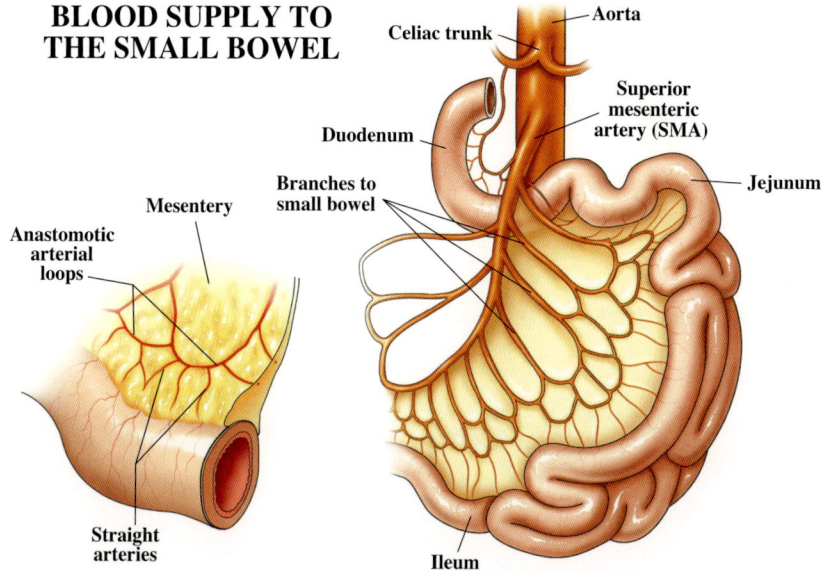

BLOOD SUPPLY TO THE SMALL BOWEL

- Aorta
- Celiac trunk
- Duodenum
- Superior mesenteric artery (SMA)
- Jejunum
- Branches to small bowel
- Mesentery
- Anastomotic arterial loops
- Straight arteries
- Ileum

GI5: Blood supply to small bowel

- With the exception of a portion of the first part of the duodenum, the small bowel is supplied by the many branches of the superior mesenteric artery.

- The branches anastomose with each other in two layers of arcades or arches, and from these, small straight vessels pass to the bowel surface, traveling around and through the wall, dividing into smaller and smaller branches.

- The arcades and multiple straight vessels are an adaptation which protect the bowel. Damage can occur to a portion of the small bowel without loss of the entire organ. Clots and ischemia from atherosclerosis and other vascular pathologies can affect the small bowel, much like the brain, heart, kidney and other organs can be affected by such conditions.

27

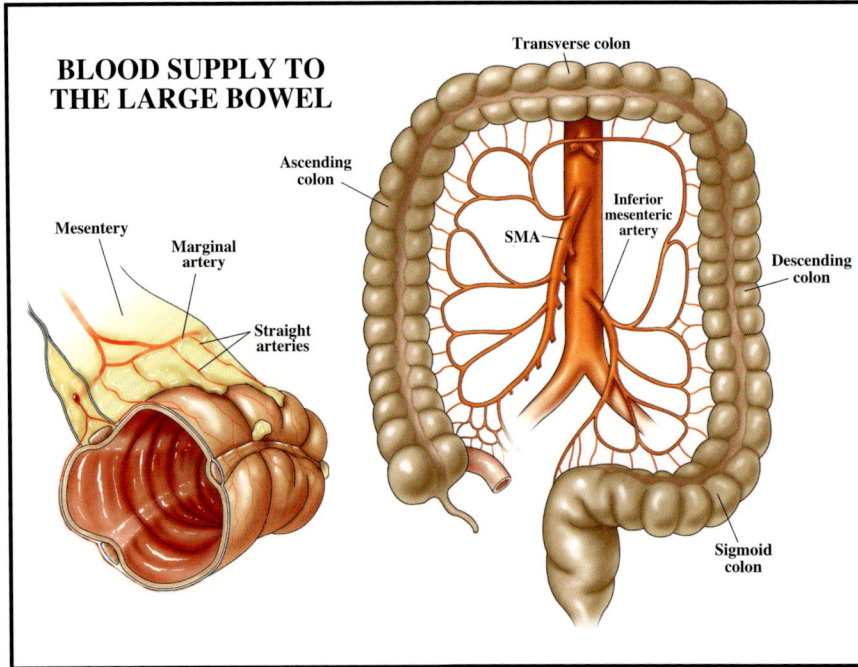

BLOOD SUPPLY TO THE LARGE BOWEL

Transverse colon

Ascending colon

Mesentery

Marginal artery

Straight arteries

SMA

Inferior mesenteric artery

Descending colon

Sigmoid colon

GI6: Blood supply to large bowel

• The blood supply of the colon comes from three sources: the superior mesenteric arteries supplying the cecum, ascending (right) colon and half of the transverse colon; the inferior mesenteric arteries supplying the distal half of the transverse colon, the descending (left) colon and the sigmoid colon; the rectal arteries supply the rectum.

• The arteries then divide into arcades, as they do to the small bowel, with straight arteries entering the bowel wall at the mesenteric border.

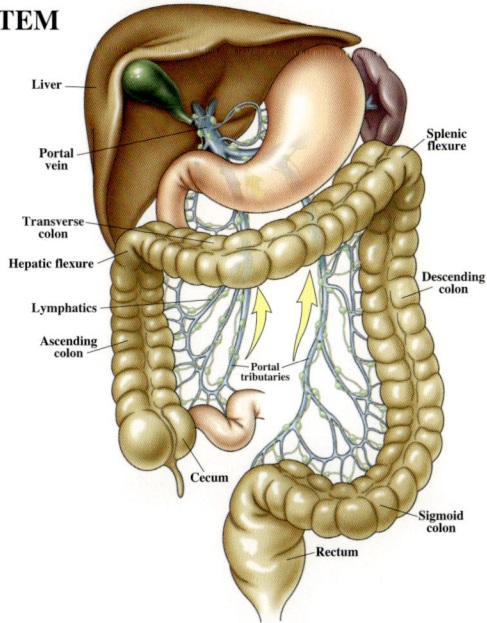

PORTAL SYSTEM

Liver

Portal vein

Transverse colon

Hepatic flexure

Lymphatics

Ascending colon

Cecum

Splenic flexure

Descending colon

Portal tributaries

Sigmoid colon

Rectum

GI7: Portal system

• The portal system is a specialized venous drainage system of the large bowel. Instead of merely draining deoxygenated blood, the portal system drains metabolites and nutrients upward so that they detour through the liver instead of returning directly to the heart and lungs. The liver serves as a cleaning and metabolic sieve where drugs and other chemicals are further broken down and either used or removed from the system.

ANATOMY

Inferior vena cava Peritoneum

Left kidney

Duodenum

Aorta

Fat

Pancreas

Ureter

Iliac vessels

Plane of section

Right kidney

BACK

Left kidney

Spine

Liver

Pancreas

Descending colon

Transverse colon

FRONT

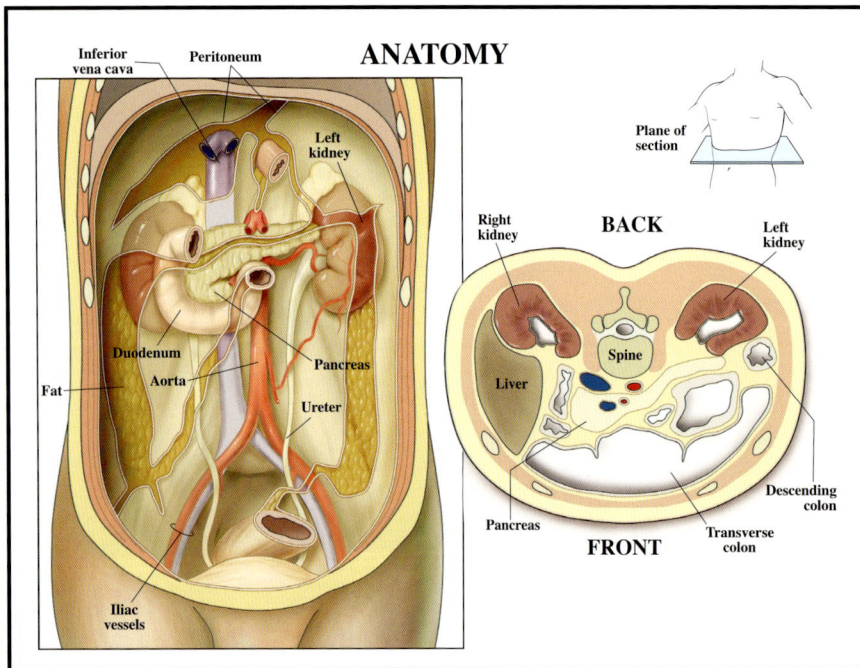

GI8: Retroperitoneum

- Most of the abdominal contents lie within the peritoneum, a sac made up of a sheet of dense connective tissue. Some structures lie behind the peritoneum (retroperitoneal). Others go in and out of it, although edges are sealed and there is little or no direct connection between the intra- and retroperitoneal regions.

- The liver has a "bare area" at its top where it lies directly against the lower surface of the diaphragm, but the rest of it is intraperitoneal. The ascending and descending colons are both retroperitoneal, while the transverse colon and part of the sigmoid are intraperitoneal; the duodenum, or first portion of the small intestine, is retroperitoneal.

- True retroperitoneal structures include the pancreas, the kidneys, ureters and adrenals, the great vessels and the pelvic structures.

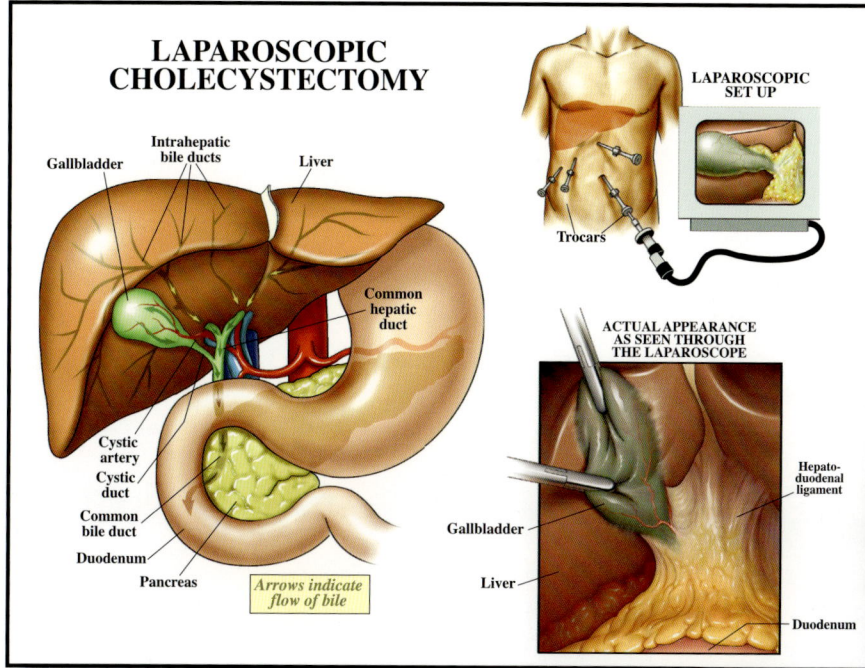

LAPAROSCOPIC CHOLECYSTECTOMY

Gallbladder

Intrahepatic bile ducts

Liver

Common hepatic duct

Cystic artery

Cystic duct

Common bile duct

Duodenum

Pancreas

Arrows indicate flow of bile

LAPAROSCOPIC SET UP

Trocars

ACTUAL APPEARANCE AS SEEN THROUGH THE LAPAROSCOPE

Hepato-duodenal ligament

Gallbladder

Liver

Duodenum

GI9: Laparoscopic cholecystectomy: Surgical set-up

- Cholecystitis, or inflammation of the gallbladder, is usually caused by gallstones blocking the cystic duct. Removal is usually performed via a laparoscopic approach, using an endoscope for visualization and hollow trocars holding the small instruments used for the surgery.

- The view through the laparoscope is transmitted to a video monitor, and the physician controls the progress by either looking directly through the scope or at the video display, depending on his or her preference and training.

- The overall complication rate for the laparoscopic procedure is about half that of the open procedure, although converting a laparoscopic procedure to an open one occurs approximately 4% of the time, usually because of difficulty in visualization.

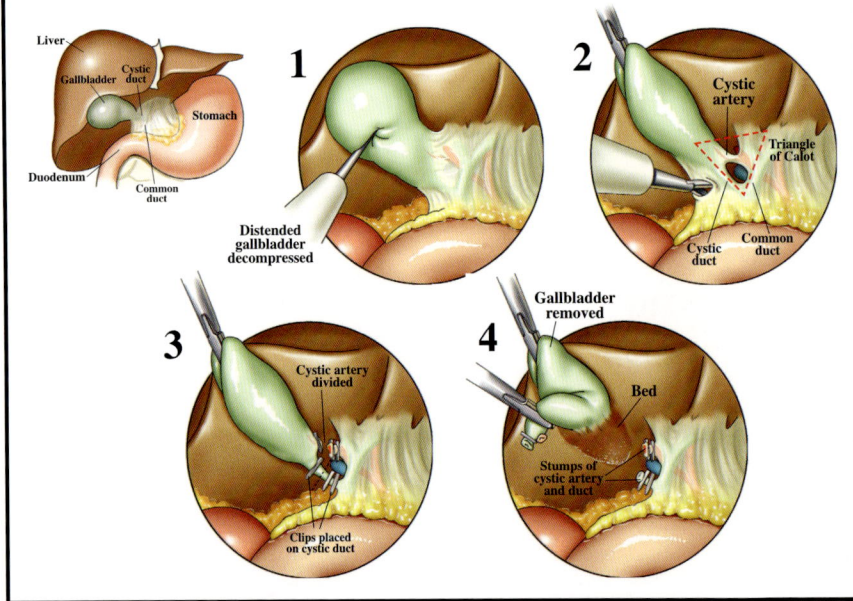

LAPAROSCOPIC CHOLECYSTECTOMY

Liver, Gallbladder, Cystic duct, Stomach, Duodenum, Common duct

1 — Distended gallbladder decompressed

2 — Cystic artery, Triangle of Calot, Cystic duct, Common duct

3 — Cystic artery divided, Clips placed on cystic duct

4 — Gallbladder removed, Bed, Stumps of cystic artery and duct

GI10: Laparoscopic cholecystectomy: Procedure

• After placement of the trocars, the gallbladder is grasped and retracted upward and outward. Adhesions, connective tissue and the lesser omentum are divided from the neck of the gallbladder in a medial direction, to reveal a portion of the cystic duct.

• Clips are placed on the exposed cystic duct and an incision is made between the clips.

• The cystic artery is then located within the Triangle of Calot (formed by the planes of the lower border of the liver, the cystic duct and the common hepatic duct), ligated and divided.

• The gallbladder is removed through one of the ports.

CHOLANGIOGRAPHY

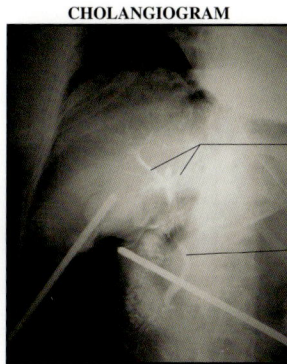

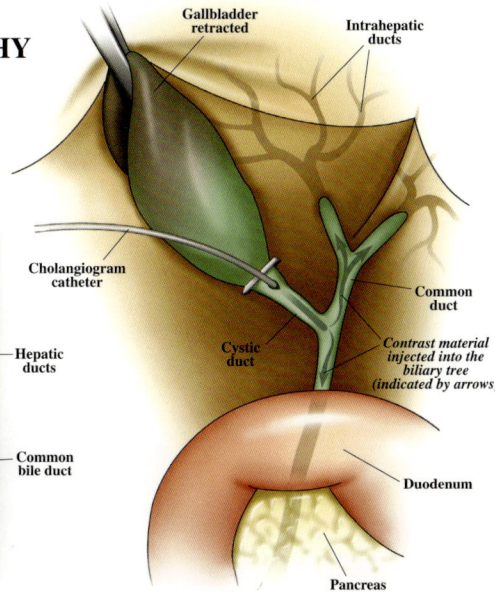

Gallbladder retracted

Intrahepatic ducts

CHOLANGIOGRAM

Cholangiogram catheter

Common duct

Hepatic ducts

Cystic duct

Contrast material injected into the biliary tree (indicated by arrows)

Common bile duct

Duodenum

Pancreas

GI11: Cholangiography

• Performed during surgery for gallbladder removal, this is an effective intraoperative radiographic test to look for either blockage or leakage in the biliary tree.

• This test may be performed prior to removing the gallbladder, or at any time a problem is suspected. A tiny catheter is threaded through a small incision in the cystic duct. Dye is injected into the biliary tract and x-rays are taken, allowing the surgeon to see which ducts are patent. Voids represent stones or tumors, and extravasation represents a leak in the system.

• While this test is very reliable in the case of a retained stone or suspected damage, outcome in patients having routine intraoperative cholangiography without apparent complication is the same as those in whom the test was not performed.

ILEUS

NORMAL
BOWEL

Wavelike contraction
of bowel walls keeps
food moving forward

ILEUS

Bowel walls
relax and distend

Air
pocket

Air pockets appear on
x-ray and abdomen is
tender and swollen

GI12: Ileus

- Normal small bowel function is in the form of peristalsis, regular wave-like contractions of the smooth muscle within the wall of the bowel. Digested food materials (chyme) move through the small bowel, where intestinal villi absorb nutrients. These raw materials enter the bloodstream and are distributed throughout the body for growth and maintenance.

- Ileus is a temporary reduction or cessation of peristalsis, allowing fluid, chyme and gases to accumulate. It is characterized by abdominal distension and discomfort; on x-ray, distended bowel loops with air/fluid levels can be seen. Bowel sounds are reduced or absent, and gas and stool are not passed.

- Ileus is a common sequela of abdominal or pelvic surgery, lasting hours to days. Symptoms are relieved by nasogastric suction to reduce pressure.

ADHESIONS

NORMAL ANATOMY
(APPEARANCE BEFORE MANIPULATION)

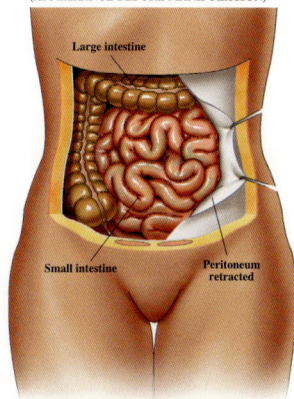

Large intestine

Small intestine

Peritoneum
retracted

DEVELOPMENT OF ADHESIONS

THIN, FILMY ADHESIONS
(Hours to days after surgery)

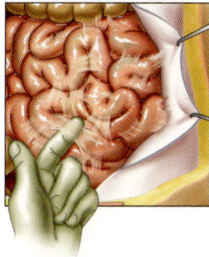

DENSER ADHESIONS
(Days to weeks after surgery)

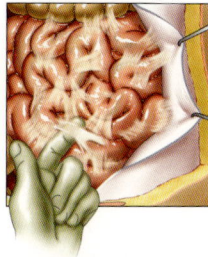

TOUGH, FIBROUS ADHESIONS
(Weeks to months after surgery)

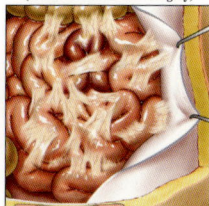

MATTING ADHESIONS
(Months to years after surgery)

GI13: Adhesions

- Adhesions are fibrous scars which can form after any disturbance within body cavities and spaces. Inciting events can include surgery, trauma and inflammation.

- Within hours of disturbance, thin, filmy strands form between bowel loops and between bowel and peritoneum and/or body wall. These continue to form for a period of time and mature over a period of weeks.

- Mature adhesions are dense, white fibrous tissues which have merged with the outer layer of the tissues; they eventually develop their own blood supply and may become severe enough to cause chronic pain and pose a chronic risk of small bowel obstruction or volvulus.

GASTROINTESTINAL

LAPAROSCOPIC GASTRIC BYPASS

Stomach

Scope

TROCAR PLACEMENT

A

Gastric pouch created

Duodenum

Stomach

Jejunum

B

Jejunum divided

C

* Arrows indicate flow of food

Jejunal limb brought up and anastomosed to gastric pouch

D Proximal jejunal stump anastomosed to jejunal limb

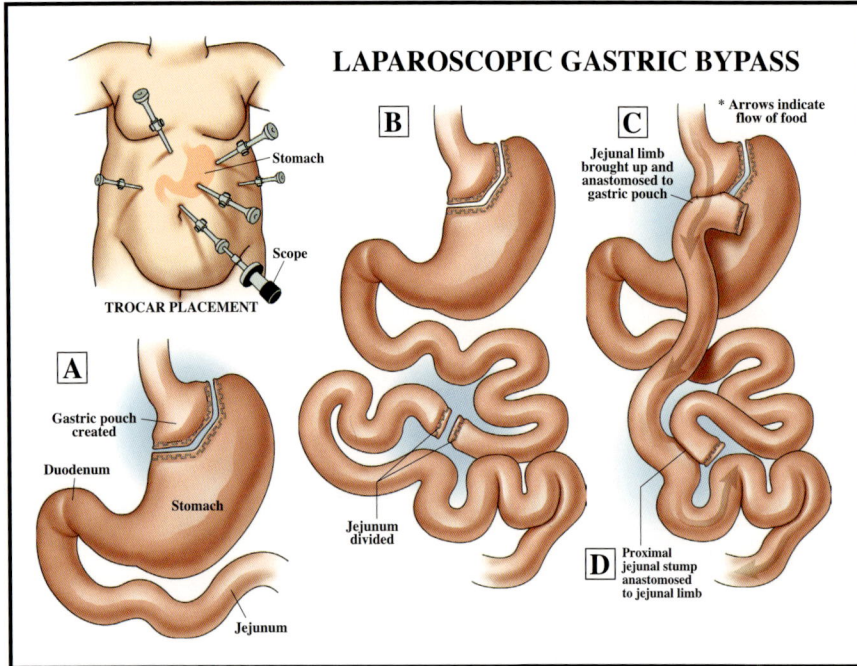

GI14 Laparoscopic gastric bypass

• Gastric bypass is performed to reduce the volume of food which the stomach can hold, and to reduce the amount of bowel available to absorb nutrients.

• There are several surgical variations; in this version, the stomach is divided and the small bowel is surgically joined to the small stomach remnant, bypassing the rest of the stomach. A second surgical anastomosis is made further down the length of the small bowel. No tissue is removed.

• The procedure can be performed either through a large abdominal incision, or laparoscopically, using "band-aid" incisions. A lighted scope is inserted into the abdomen, as are several slender tubes. Instrumentation is then placed into the tubes and the procedure is performed under direct vision through the scope.

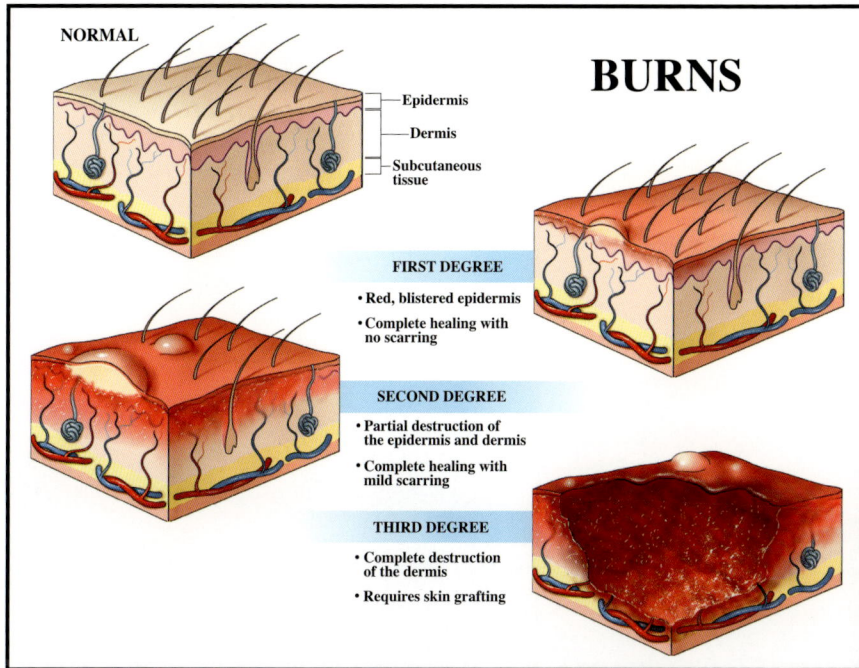

NORMAL

Epidermis

Dermis

Subcutaneous tissue

BURNS

FIRST DEGREE
- Red, blistered epidermis
- Complete healing with no scarring

SECOND DEGREE
- Partial destruction of the epidermis and dermis
- Complete healing with mild scarring

THIRD DEGREE
- Complete destruction of the dermis
- Requires skin grafting

I1: Burns

- The skin is a very large, thin structure covering the entire body. All skin, with the exception of the palms and soles, bears hair. It is a rapidly-regenerating structure.

- The epidermis has layered keratinized cells; the bottom-most layer of the epidermal cells is constantly dividing. The dermis, which lies under the epidermis, contains blood vessels, hair follicles, sweat and oil glands and several kinds of nerve fibers. Beneath the dermis is fatty subcutaneous tissue.

- Burns are classified by their depth. First degree burns result in a reddened, sore epidermis. Second degree burns extend partially through the dermis and usually result in severe pain and blistering. Third degree burns, which go completely through the dermis, are relatively pain free because the nerve fibers are gone. Fluid loss is a serious problem for patients with 3rd degree burns.

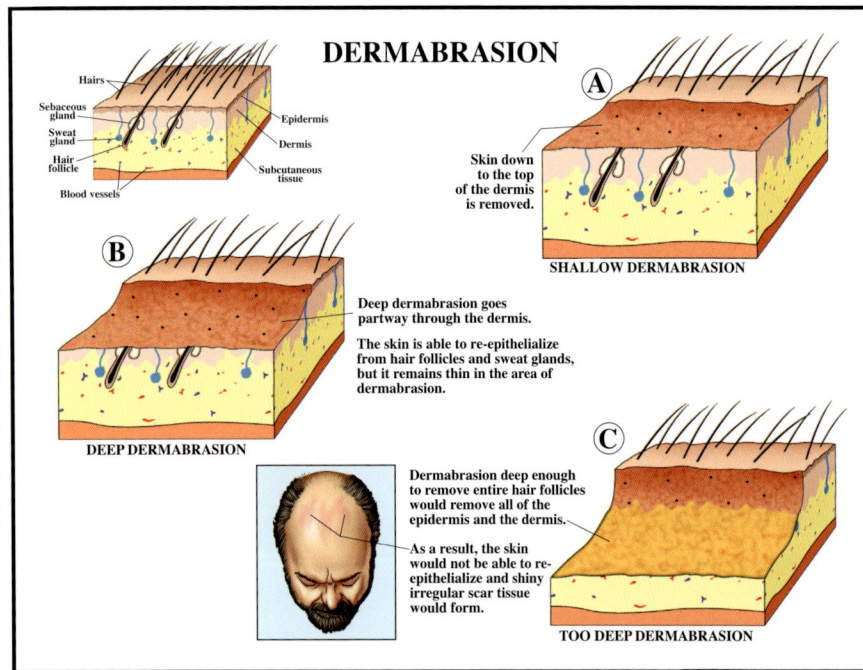

DERMABRASION

Hairs

Sebaceous gland

Sweat gland

Hair follicle

Blood vessels

Epidermis

Dermis

Subcutaneous tissue

Ⓐ

Skin down to the top of the dermis is removed.

SHALLOW DERMABRASION

Ⓑ

Deep dermabrasion goes partway through the dermis.

The skin is able to re-epithelialize from hair follicles and sweat glands, but it remains thin in the area of dermabrasion.

DEEP DERMABRASION

Ⓒ

Dermabrasion deep enough to remove entire hair follicles would remove all of the epidermis and the dermis.

As a result, the skin would not be able to re-epithelialize and shiny irregular scar tissue would form.

TOO DEEP DERMABRASION

I2: Dermabrasion

• This is a common technique for smoothing skin scarred from acne or trauma, or for reducing small wrinkles. A wire brush or laser is used to remove the top layer of skin as evenly as possible. The epidermis is removed down to the top portion of the dermis, which then regenerates in a smooth layer. Large wrinkles and deep scars can be improved but will not completely disappear.

• Deep dermabrasion removes the skin down further into the dermis; skin will be somewhat smoother, but thinner than untreated skin.

• If nearly the entire thickness of the dermis is removed, the epidermis will regenerate, but hair will not grow, the glands will not regenerate, and the skin will have a smooth, shiny, unnatural appearance.

SPLIT-THICKNESS SKIN GRAFTING

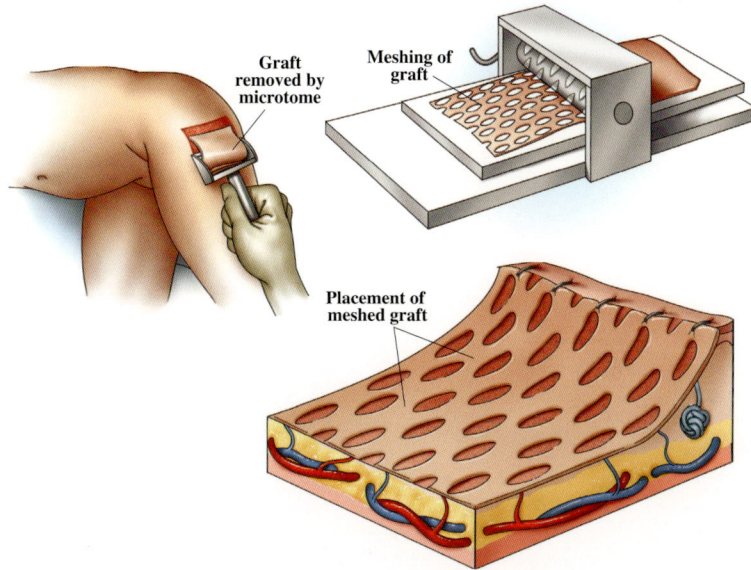

Graft removed by microtome

Meshing of graft

Placement of meshed graft

I3: Split-thickness skin grafting

- When full-thickness skin injuries occur—all dermis is gone to the level of the subcutaneous tissue or below—skin grafting is required to cover the defect and stop fluid leakage.

- Split-thickness skin grafting (STSG) involves removal of thin strips of healthy skin down to the dermis. The strips are then put through a meshing machine which significantly increases the total area of coverage; the meshed portions fill in with healing.

- STSG does not require microanastomosis of blood vessels because oxygen, glucose, protein and other reparative raw materials can diffuse from the donor site vessels to keep it alive until the vasculature is fully healed. Artificial skin, pig skin and donor skin are also used to cover and protect severely injured areas of skin.

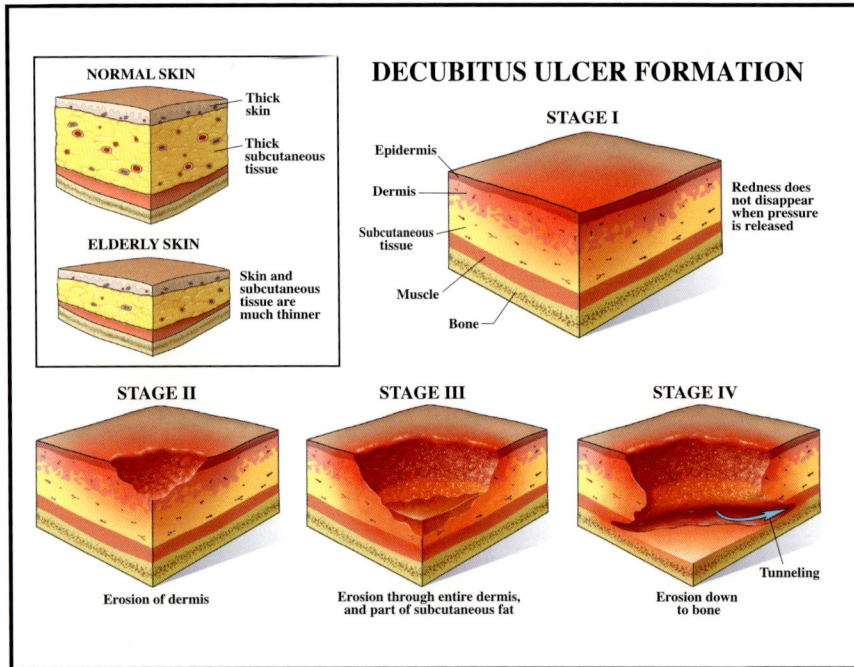

DECUBITUS ULCER FORMATION

NORMAL SKIN
- Thick skin
- Thick subcutaneous tissue

ELDERLY SKIN
- Skin and subcutaneous tissue are much thinner

STAGE I
- Epidermis
- Dermis
- Subcutaneous tissue
- Muscle
- Bone
- Redness does not disappear when pressure is released

STAGE II
Erosion of dermis

STAGE III
Erosion through entire dermis, and part of subcutaneous fat

STAGE IV
- Tunneling
Erosion down to bone

I4: Decubitus ulcer formation

- Decubitus ulcers are common in debilitated patients, particularly the aged.

- Most decubitus patients have thin skin with thin subcutaneous tissue, but poorly-vascularized fatty tissue can also predispose a vulnerable patient to develop these lesions.

- The stages are similar to those of burns, although decubiti are caused by pressure rather than temperature. Stage IV decubiti are full-thickness defects and are characterized by undermining; they are frequently larger beneath the skin than on the surface.

- In a debilitated patient, decubiti can form in as little as 15 minutes. Healing is difficult, usually due to widespread vascular disease, nutritional deficits and diabetes.

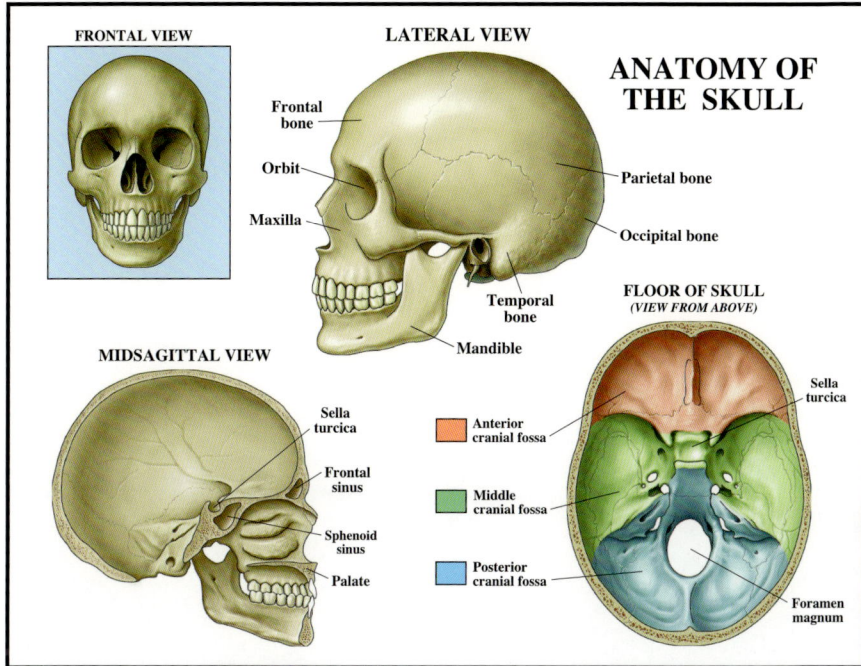

FRONTAL VIEW

LATERAL VIEW

ANATOMY OF
THE SKULL

Frontal
bone

Orbit

Maxilla

Parietal bone

Occipital bone

Temporal
bone

Mandible

MIDSAGITTAL VIEW

FLOOR OF SKULL
(VIEW FROM ABOVE)

Sella
turcica

Frontal
sinus

Sphenoid
sinus

Palate

Anterior
cranial fossa

Middle
cranial fossa

Posterior
cranial fossa

Sella
turcica

Foramen
magnum

M1: Anatomy of the skull

• The calvarium, which houses the brain, is composed of several large, flat bones. The sutures between the calvarial bones are strong, with interdigitating fingers of bone at the joints; the frontal bone consists of two separate bones that fuse by the age of 18 years.

• The base of the skull has three sections, each corresponding to a major part of the brain (anterior cranial fossa for the frontal lobes, middle cranial fossa for the temporal lobes, and posterior cranial fossa for the cerebellum). Cranial nerves and vessels enter and exit through openings in the floor of the skull.

• The bones of the face have air-filled cavities or sinuses to make them lighter and to help warm inhaled air. The only moveable joint is the bilateral temporomandibular joint between the mandible and calvarium.

MUSCULOSKELETAL

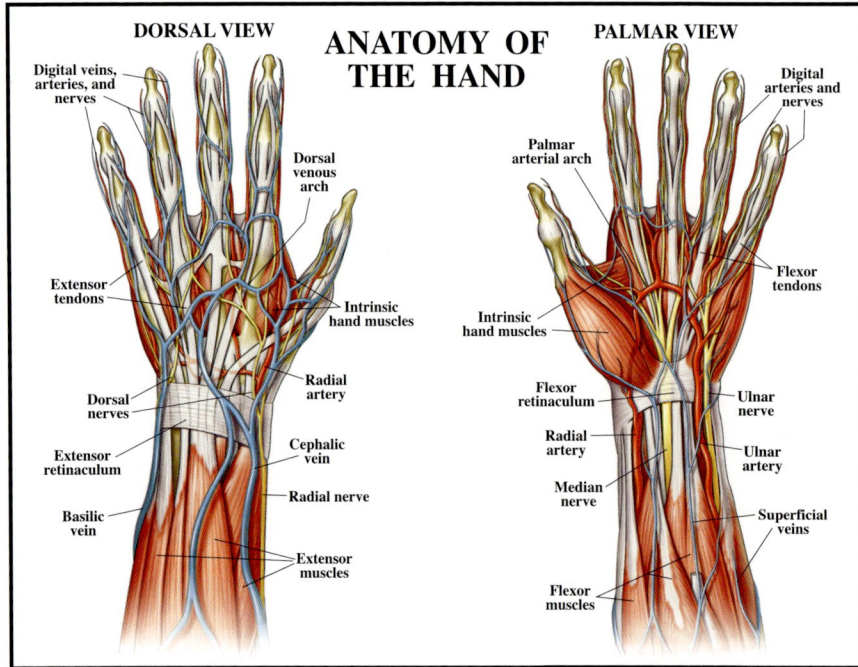

DORSAL VIEW

ANATOMY OF THE HAND

PALMAR VIEW

Digital veins, arteries, and nerves

Dorsal venous arch

Digital arteries and nerves

Palmar arterial arch

Extensor tendons

Intrinsic hand muscles

Flexor tendons

Intrinsic hand muscles

Dorsal nerves

Radial artery

Flexor retinaculum

Ulnar nerve

Extensor retinaculum

Cephalic vein

Radial artery

Ulnar artery

Basilic vein

Radial nerve

Median nerve

Superficial veins

Extensor muscles

Flexor muscles

M2: Anatomy of the hand

- One of the more complex structures, the hand has a high concentration of nerves and vessels. There are also many small intrinsic muscles which allow fine motor function.

- Tendons, blood vessels and nerves originating in the arm cross the wrist to enter the hand; the intrinsic muscles are solely within the hand.

- The structures crossing the double row of wrist bones are held in place by the flexor retinaculum on the volar (palmar) side ("carpal tunnel"), and by the extensor retinaculum on the dorsal side. The flexor retinaculum can sometimes thicken or scar, causing compression of the median nerve, or carpal tunnel syndrome.

NORMAL KNEE ANATOMY

VIEW OF KNEE IN FLEXION

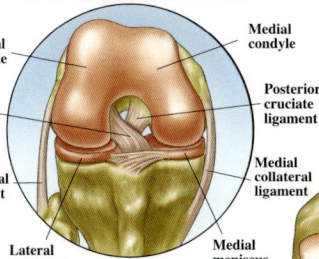

Lateral condyle
Medial condyle
Anterior cruciate ligament
Posterior cruciate ligament
Lateral collateral ligament
Medial collateral ligament
Lateral meniscus
Medial meniscus

Femur
Patella
Fibula
Tibia

FRONT VIEW

TOP VIEW

Lateral meniscus
Medial meniscus
Anterior cruciate ligament
Posterior cruciate ligament

Femur
Posterior cruciate ligament
Medial condyle
Lateral condyle
Fibula

BACK VIEW

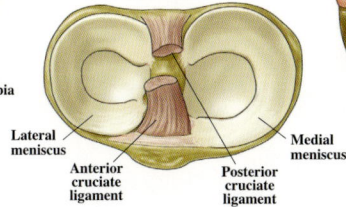

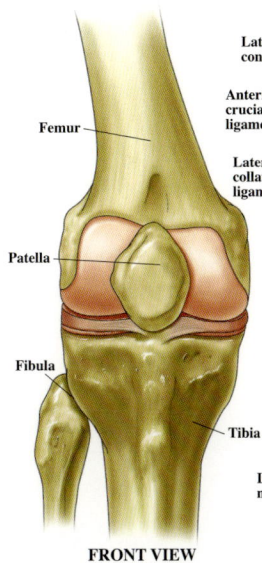

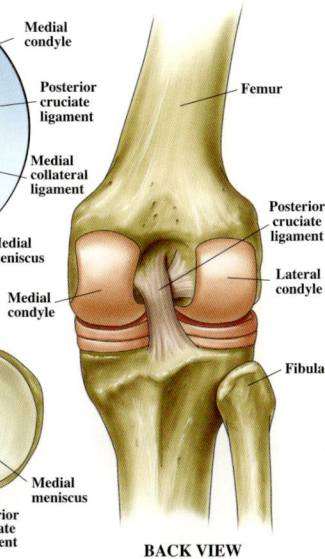

M3: Normal knee anatomy

• One of the major joints in the body, the knee is required to support huge and repeated pressures over the course of a lifetime. The articular surfaces of the joint are covered with glassy-smooth articular cartilage, and the menisci act as shock absorbers with each step.

• The patella is a large sesamoid bone which lies within the quadriceps tendon. It articulates with the condyles of the femur.

• The anterior and posterior cruciate ligaments are located at the center of the joint and allow some rotatory motion. The lateral and medial collateral ligaments are attached on either side of the joint to maintain stability.

NORMAL SHOULDER ANATOMY

Coracoclavicular ligament
Clavicle
Acromioclavicular joint
Acromion
Coracoacromial ligament
Coracoid process
Humerus
Scapula
Subscapularis muscle

ANTERIOR VIEW

SIDE VIEW OF JOINT

Acromion
Supraspinatous tendon
Infraspinatous tendon
Glenoid fossa
Teres minor tendon
Subscapularis tendon
Middle glenohumeral ligament
Joint capsule

Clavicle
Spine of scapula
Supraspinatus muscle
Acromion
Humerus
Teres minor muscle
Scapula
Infraspinatus muscle

POSTERIOR VIEW

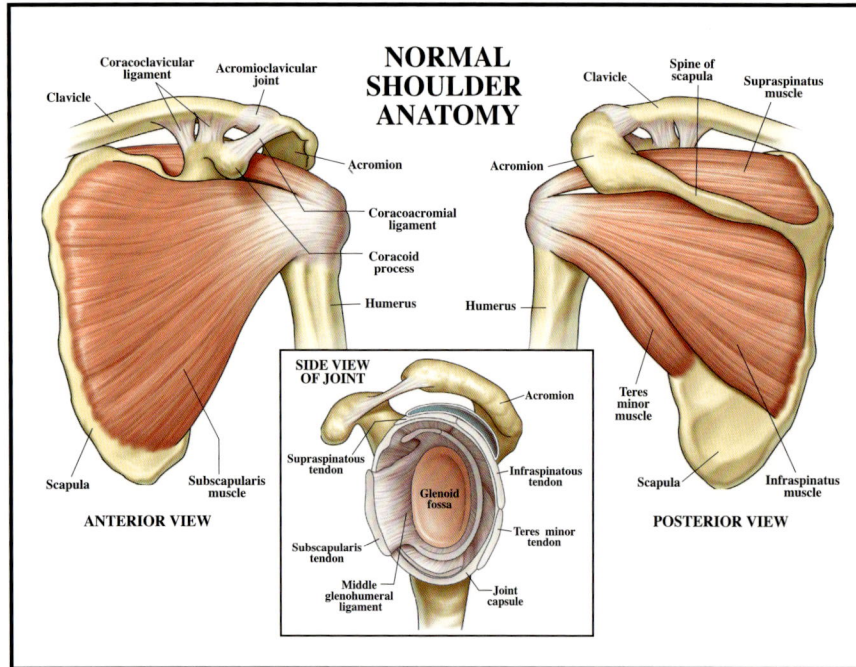

M4: Normal shoulder anatomy

- The tendons of the deep shoulder muscles (infraspinatus, supraspinatus and teres minor) conjoin with the shoulder joint capsule to form the rotator cuff.

- Rotator cuff injuries are common and may be difficult to treat.

- The head of the humerus rests in the glenoid fossa, a relatively shallow depression in the scapula. The rotator cuff holds the head in the fossa during movement.

- The shoulder joint is prone to dislocation due to the shallowness of the glenoid fossa.

- The acromioclavicular joint lies over the shoulder and sometimes develops fibrosis, making movement painful or stiff.

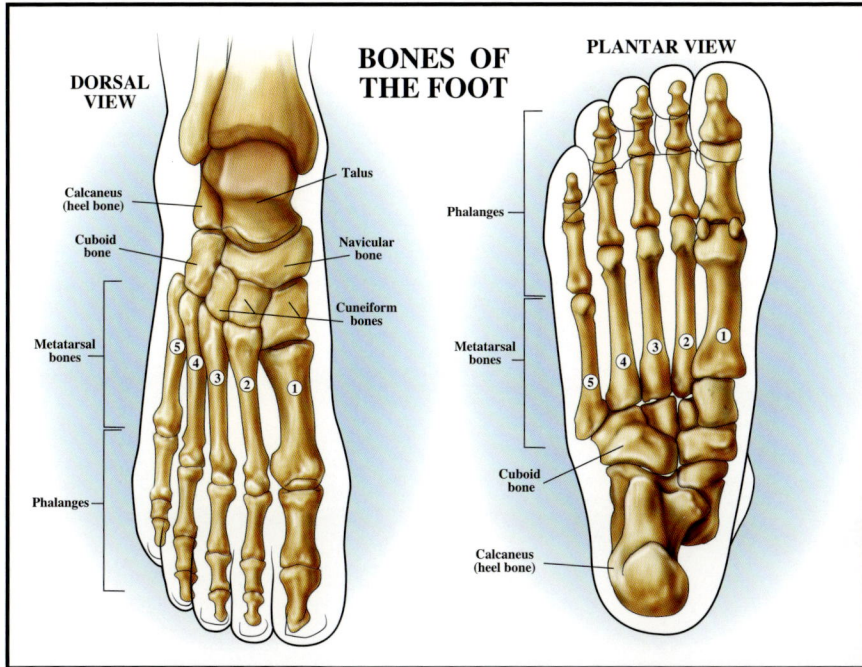

BONES OF THE FOOT

DORSAL VIEW

- Calcaneus (heel bone)
- Cuboid bone
- Metatarsal bones
- Phalanges
- Talus
- Navicular bone
- Cuneiform bones
- 5 4 3 2 1

PLANTAR VIEW

- Phalanges
- Metatarsal bones
- 5 4 3 2 1
- Cuboid bone
- Calcaneus (heel bone)

M5: Bones of the foot

- The bones of the foot follow basically the same pattern of the hand bones. There is a double layer of sesamoid-like bones forming the ankle, which articulate with the long bones making up the central foot (metatarsals), which are in turn attached to the phalanges, or toes.

- The tendons and the many ligaments of the foot attach to the tough, thin tissue covering the bones, the periosteum. The ligaments attach the bones to each other, and the tendons connect the muscles to the bones.

- As in the hand, there are intrinsic muscles in the feet.

- The bones of the foot form a longitudinal arch and a transverse arch. Most of the body's weight is borne on the metatarsal heads, particularly the first and fifth.

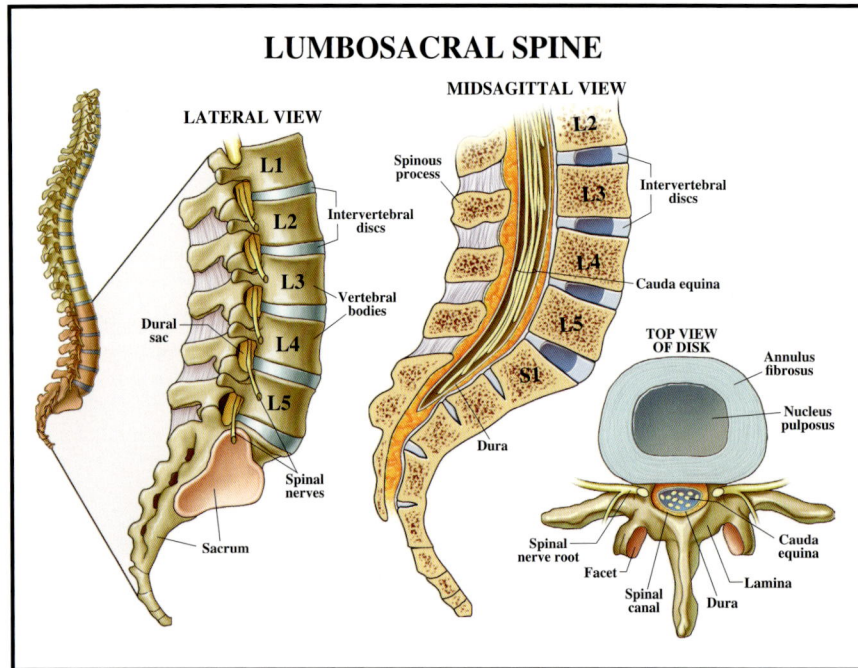

LUMBOSACRAL SPINE

LATERAL VIEW

L1
L2
L3
L4
L5

Intervertebral discs

Vertebral bodies

Dural sac

Spinal nerves

Sacrum

MIDSAGITTAL VIEW

L2
L3
L4
L5
S1

Spinous process

Intervertebral discs

Cauda equina

Dura

TOP VIEW OF DISK

Annulus fibrosus

Nucleus pulposus

Spinal nerve root

Facet

Spinal canal

Dura

Cauda equina

Lamina

M6: Lumbosacral spine

- The lumbar spine is composed of large, strong bones which must support the entire weight of the spine and head.

- The vertebral bodies are separated by fibrous discs which serve as shock absorbers. The discs have a fibrous ring (annulus fibrosis) and a gel-like center (nucleus pulposus).

- The spinal canal is formed by the pedicles, laminae, and the vertebral bodies and discs; the canal protects the distal portion of the axial nervous system, the cauda equina.

- The large transverse and spinous processes serve as support for the many paraspinous muscles which allow for the fine movement of the spine.

- The sacrum is the large, wedge-shaped bone forming the posterior part of the pelvic bowl, and is composed of fused vertebrae.

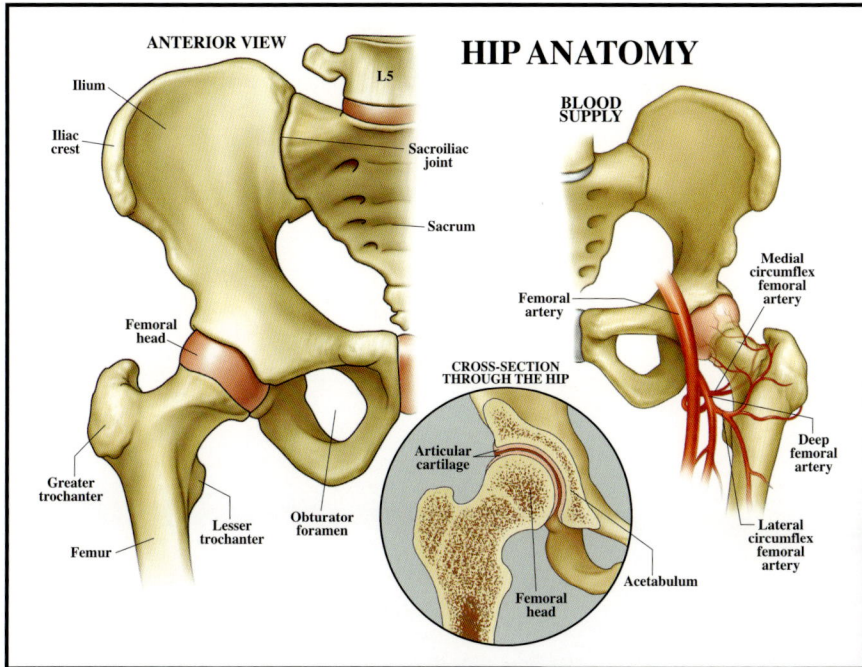

ANTERIOR VIEW

HIP ANATOMY

Ilium

Iliac crest

L5

Sacroiliac joint

Sacrum

BLOOD SUPPLY

Femoral head

Femoral artery

Medial circumflex femoral artery

Greater trochanter

Lesser trochanter

Obturator foramen

Femur

CROSS-SECTION THROUGH THE HIP

Articular cartilage

Femoral head

Acetabulum

Deep femoral artery

Lateral circumflex femoral artery

M7: Hip anatomy

• The largest joint in the body, the hip is composed of the large, round head of the femur which lies within the acetabulum or cup of the pelvis. Cartilage covers the articular surfaces, as in every other joint. There is a joint capsule and a number of muscles which cross and protect the joint and allow movement in a number of planes.

• The blood supply to the hip is relatively meager and easily disrupted with trauma.

• Since the entire weight of the body goes through this joint with every step, it is vulnerable to damage from use and is a common site for degenerative joint disease.

47

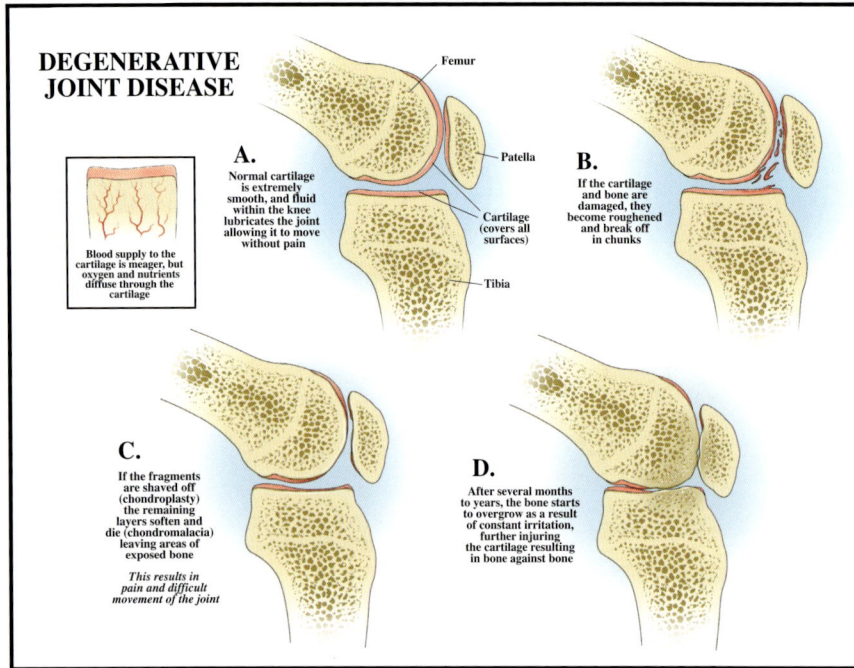

DEGENERATIVE JOINT DISEASE

Blood supply to the cartilage is meager, but oxygen and nutrients diffuse through the cartilage

Femur

Patella

Cartilage (covers all surfaces)

Tibia

A. Normal cartilage is extremely smooth, and fluid within the knee lubricates the joint allowing it to move without pain

B. If the cartilage and bone are damaged, they become roughened and break off in chunks

C. If the fragments are shaved off (chondroplasty) the remaining layers soften and die (chondromalacia) leaving areas of exposed bone

This results in pain and difficult movement of the joint

D. After several months to years, the bone starts to overgrow as a result of constant irritation, further injuring the cartilage resulting in bone against bone

M8: Degenerative joint disease

- Degenerative joint disease is a result of normal activity and is found in most people as they get older; it develops more rapidly and more severely in the case of joint trauma.

- Articular surfaces are covered with glassy-smooth cartilage. As the cartilage disintegrates over time (chondromalacia), it flakes off until bone is eventually exposed.

- If bone starts to rub against bone, it reacts by forming more bone in the form of osteophytes. This is a very painful condition and when severe enough, requires joint replacement.

- Degenerative joint disease is frequently called osteoarthritis; rheumatoid arthritis is an autoimmune disease with very different causes and a slightly different set of symptoms.

CERVICAL SPINE

MIDSAGITTAL VIEW

FRONT VIEW

C1 (Atlas)
C2 (Axis)

Nerve roots

C2
C3
C4
C5
C6
C7

C3
C4
C5
C6
C7

SPINE

C1

Intervertebral disc

Vertebrae

Spinal canal

Spinal cord

Spinous process

TOP VIEW OF C6

Foramen

Disc

Vertebral artery

LEFT

RIGHT

Nerve root

Pulposus

Spinal cord

Annulus

Spinal canal

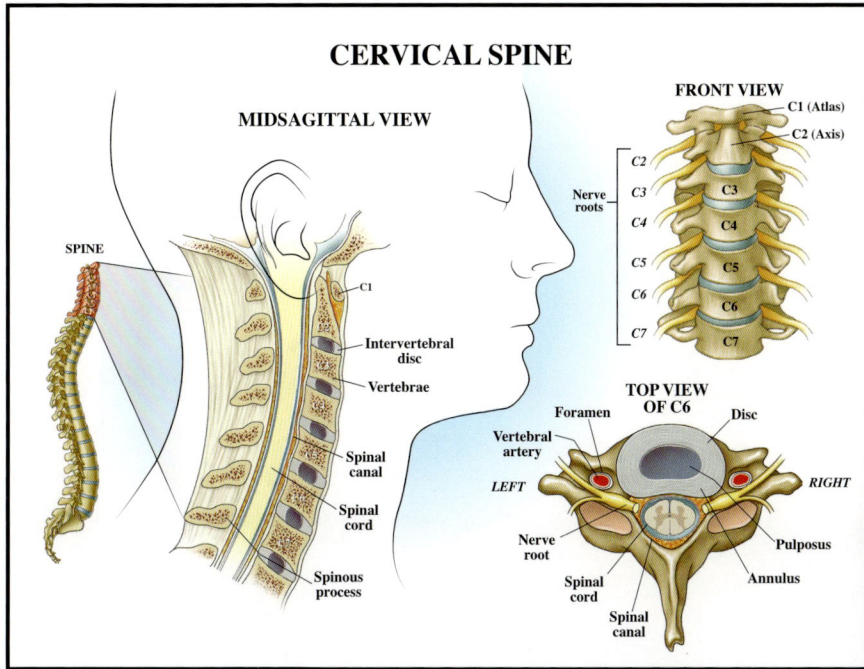

M9: Cervical spine

- The 7 vertebrae of the cervical spine help support the skull, and protect the spinal cord as it exits the cranium to pass downward through the spinal canal.

- The transverse processes each have a small hole through which the vertebral arteries pass to join to form the basilar artery supplying the posterior and deep portions of the brain and brainstem.

- The cervical nerve roots form the brachial plexus which supplies sensation and movement to the upper extremities.

- Degenerative joint disease and disc disease are very common in the cervical spine, leading to arm and hand pain and dysfunction requiring decompression and sometimes fusion.

TOTAL KNEE REPLACEMENT

A — Knee flexed
- Femur pulled anteriorly
- Cuts made in distal end of femur
- Fibula
- Tibia

B — Knee flexed
- Tibial plateau removed
- Tibia pulled anteriorly

D — Knee flexed
- Tibial component placed
- Practice components removed and permanent components placed
- Tibia pulled anteriorly

C — Knee extended to test function of components
- Patella
- Practice components placed

E — Knee extended to test function of components
- Patellar button
- Femoral component
- Tibial component

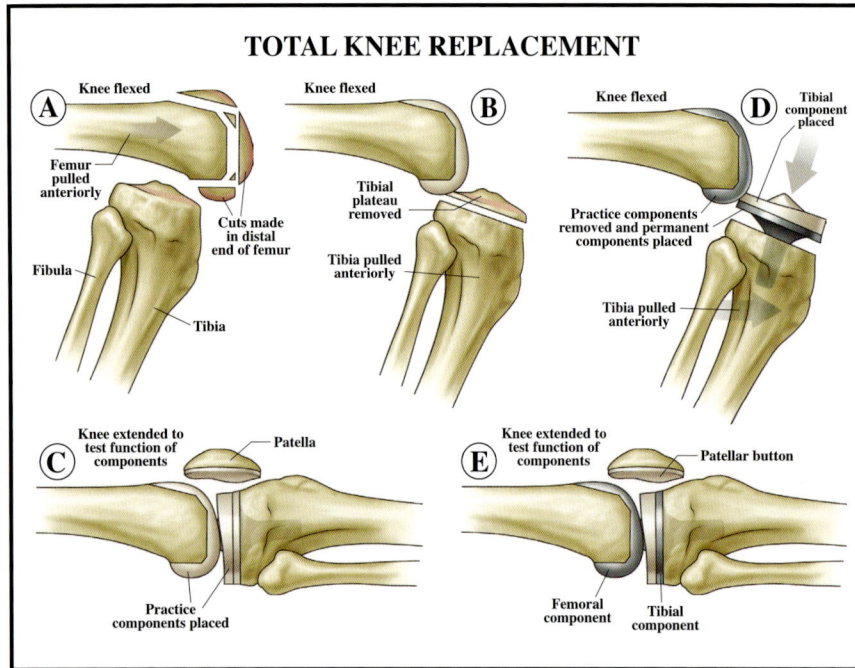

M10: Total knee replacement

- One of the more common surgical procedures, diseased bone and articular surfaces (usually from either trauma and/or degenerative joint disease) are surgically removed.

- Most knee prostheses are made of a strong metal with a durable plastic overlay to serve as a joint surface.

- As with any bony prosthesis, these can be either cemented into place, or have a roughened surface which allows bone to grow into the surface.

- All three units (femoral, tibial and patellar) can be inserted together or individually, depending upon the specific needs of the patient.

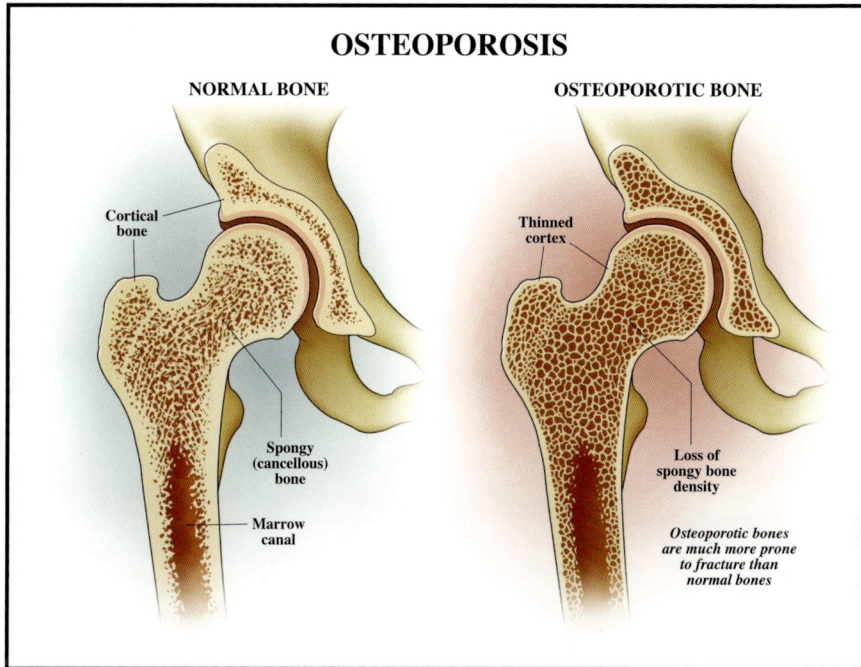

OSTEOPOROSIS

NORMAL BONE

OSTEOPOROTIC BONE

Cortical bone

Thinned cortex

Spongy (cancellous) bone

Marrow canal

Loss of spongy bone density

Osteoporotic bones are much more prone to fracture than normal bones

M11: Osteoporosis

- A common condition, particularly among elderly women, osteoporosis is the reduction of bone density.

- Osteoporosis affects both the dense cortical portion on the outside of a bone and the spongy bone inside.

- Causes include age, low body mass, family history, smoking, malnutrition, alcoholism, insufficient physical activity and certain medications and toxins. It is also associated with a number of chronic diseases.

- The primary risk is that of fracture, particularly of the hip, a potentially devastating event in the elderly. In severely osteoporotic patients, healing may be slow and insufficient.

- Treatment depends on the cause, but includes medication, dietary changes and exercise.

BRAIN SURFACE ANATOMY

LATERAL VIEW

Central sulcus
Parietal lobe
Frontal lobe
Occipital lobe
Lateral sulcus
Cerebellum
Temporal lobe
Brainstem

MID-SAGITTAL VIEW

Parietal lobe
Corpus callosum
Occipital lobe
Frontal lobe
Cerebellum
Temporal lobe
Pons
Pituitary gland
4th ventricle
Medulla oblongata

BASE OF BRAIN

Optic chiasm
Pituitary gland
Mammillary body
Pons
Brainstem
Cerebellum
I
II
III
IV
V
VI
VII
VIII
IX
X
XII
XI
Cranial nerves

POSTERIOR VIEW

Parietal lobe
Occipital lobe
Cerebellum
Brainstem

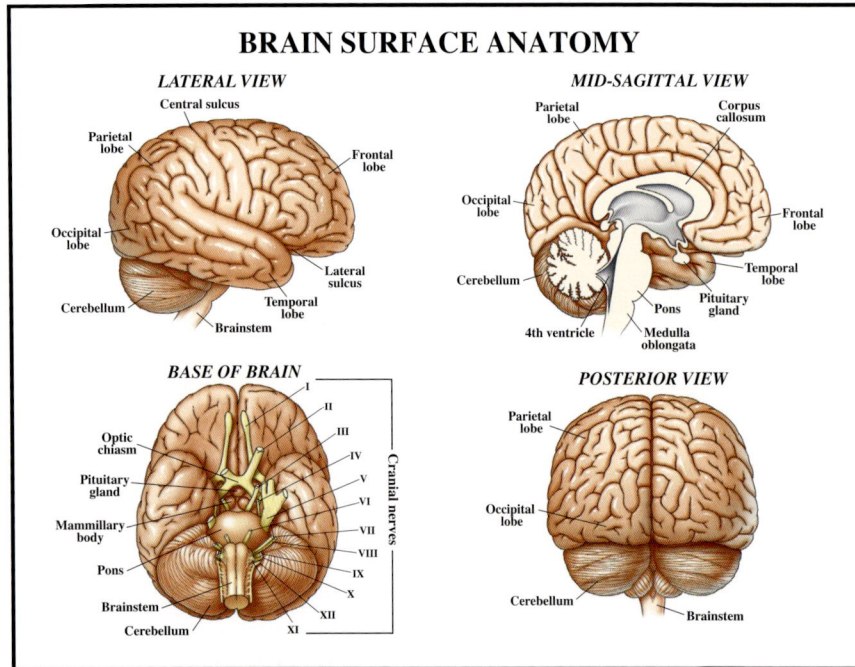

N1: Brain surface anatomy

- The surface of the brain has multiple folds, or gyri, separated by sulci. While many of these are specific to a particular individual, some are constant and serve as landmarks for functional control.

- The cerebrum is the large, rounded portion of the brain and is the site of higher functions. The cerebellum and brainstem control more basic functions like heart rate, balance and respiration.

- The cranial nerves mostly originate in the brainstem and exit the skull via foramina in the base of the skull (see M1). The optic nerves originate in the occipital lobes, travel along pathways inside the brain matter, and exit anteriorly.

- The corpus callosum is a group of myelin-covered neuronal fibers (white matter) which connect the right side of the brain with the left.

ANATOMY OF THE SPINAL CORD

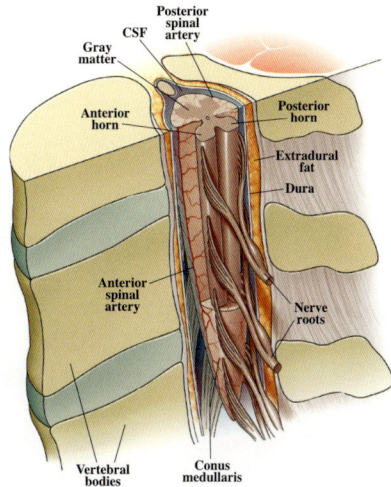

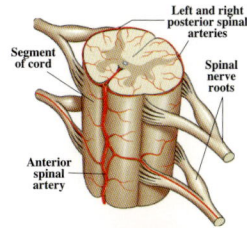

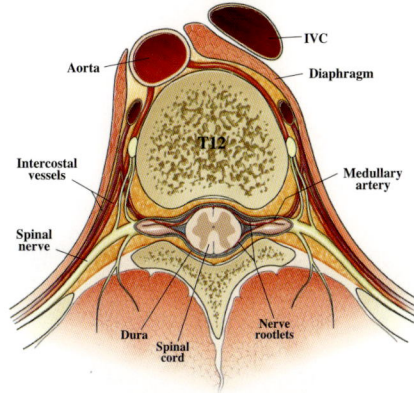

Posterior spinal artery

CSF

Gray matter

Anterior horn

Posterior horn

Extradural fat

Dura

Anterior spinal artery

Nerve roots

Vertebral bodies

Conus medullaris

Left and right posterior spinal arteries

Segment of cord

Spinal nerve roots

Anterior spinal artery

IVC

Aorta

Diaphragm

Intercostal vessels

T12

Medullary artery

Spinal nerve

Dura   Spinal cord

Nerve rootlets

N2: Anatomy of spinal cord

- The spinal cord lies within the spinal canal, formed by the vertebral bodies and the bony arch formed by the pedicles and laminae (see M6).

- The cord and its terminal nerves, the cauda equina, lie within the dural sac, a tough membranous structure filled with cerebrospinal fluid bathing and protecting the cord.

- The spinal cord itself ends at the level of the first lumbar vertebra, but nerve roots travel inside the dural sac to exit at lower levels; these roots form the cauda equina ("horse's tail").

- Blood supply comes from the segmental branches of the aorta, traveling along the nerve root to merge as the anterior spinal artery running in the front midline of the cord; there are two parallel vessels along the back surface of the cord. There is also a generous venous plexus within the canal.

53

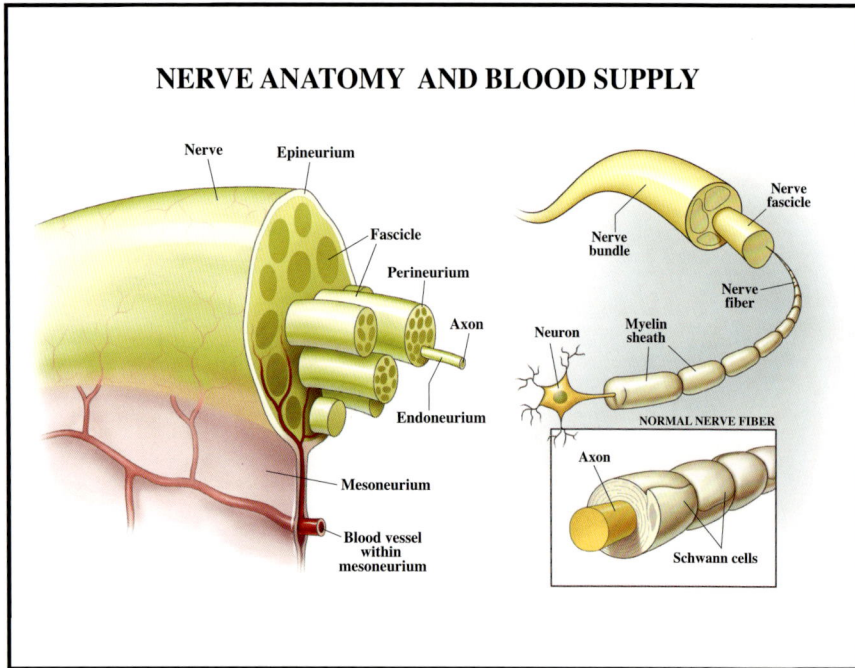

NERVE ANATOMY AND BLOOD SUPPLY

Nerve
Epineurium
Fascicle
Perineurium
Axon
Endoneurium
Mesoneurium
Blood vessel within mesoneurium

Nerve bundle
Nerve fascicle
Nerve fiber

Neuron
Myelin sheath

NORMAL NERVE FIBER
Axon
Schwann cells

N3: Nerve anatomy/blood supply

- Nerves are composed of multiple fibers which in turn are composed of many, many axons. Each successive grouping of fibers is surrounded by connective tissue.

- Axons are the long branches of nerve cells which transmit electrochemical impulses from one nerve to another or to and from target organs.

- The axons are surrounded by myelin, a kind of insulating material formed by oligodendrocytes. In diseases in which myelin is destroyed, electrochemical conduction is impossible, and even if the nerve fibers do not die, they are ineffective.

- Blood vessels travel through the mesoneurium and then divide into capillaries within the nerve to deliver oxygen and nutrients.

- Nerves are vulnerable to trauma, reduced oxygen levels and diabetes.

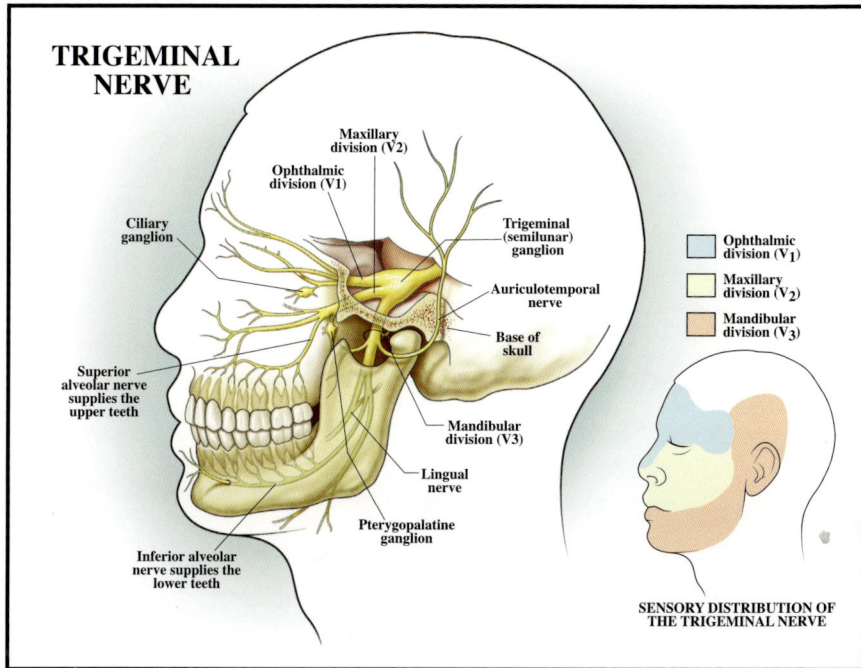

TRIGEMINAL NERVE

Maxillary division (V2)

Ophthalmic division (V1)

Ciliary ganglion

Trigeminal (semilunar) ganglion

Auriculotemporal nerve

Base of skull

Superior alveolar nerve supplies the upper teeth

Mandibular division (V3)

Lingual nerve

Pterygopalatine ganglion

Inferior alveolar nerve supplies the lower teeth

	Ophthalmic division (V$_1$)
	Maxillary division (V$_2$)
	Mandibular division (V$_3$)

SENSORY DISTRIBUTION OF THE TRIGEMINAL NERVE

N4: Trigeminal nerve

- The semilunar ganglion of the trigeminal nerve lies deep within the skull. The three branches of the nerve leave through large separate openings in the base of the skull.

- The trigeminal nerve is the 5th cranial nerve. Its three portions are the ophthalmic nerve controlling sensation to the upper face, the maxillary nerve controlling sensation to the mid-face, and the mandibular nerve supplying sensation to the lower face and the skin around and above the ear. Each of the branches supplies both soft tissue and bone.

- All large ganglia in the body can harbor certain viruses. Oral herpes infections "hibernate" within the semilunar ganglion, traveling down the nerve roots to the mouth when the virus is activated by stress.

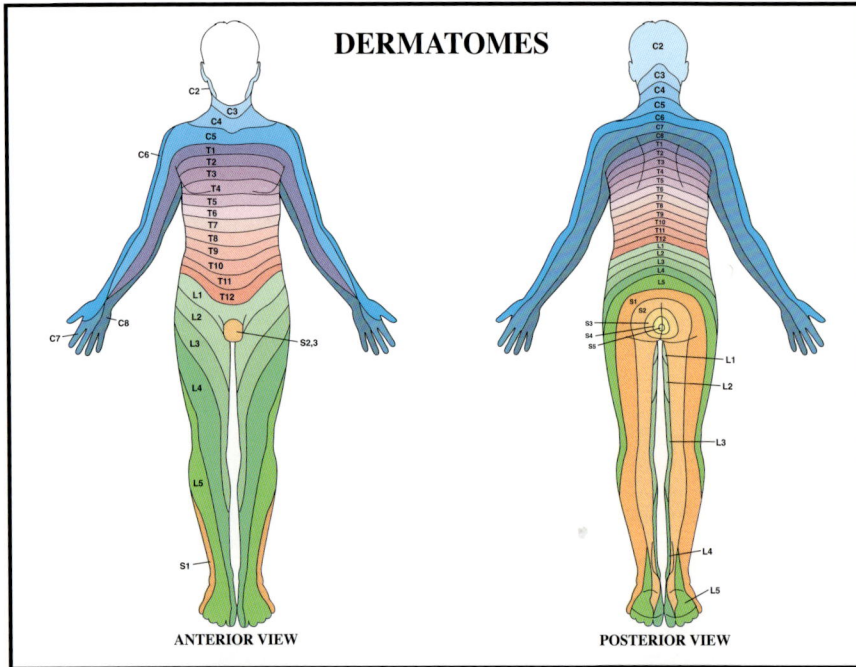

DERMATOMES

ANTERIOR VIEW

POSTERIOR VIEW

N5: Dermatomes

• Dermatomes are strips of skin which are supplied by the nerve roots. If there is numbness or pain along a dermatome, it is a sign of damage or irritation of a specific nerve root, where the root exits the spinal cord and vertebral column. This is known as radiculopathy.

• Soon after the cervical and lumbosacral nerves leave the cord, they join and separate several times (plexi) before reaching their target organs. Nerve fibers to a given muscle may come from several different nerve roots. The skin sensory supply, however, remains directly associated with the root alone.

• Radicular pain usually occurs with compression of the nerve in the foramen, the hole by which the nerve exits the spinal canal.

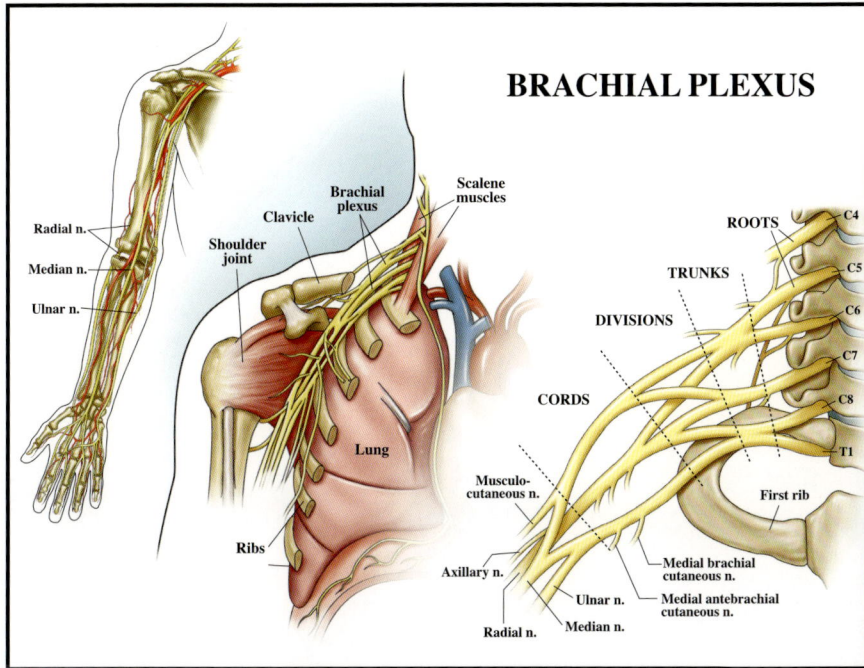

BRACHIAL PLEXUS

Radial n.

Median n.

Ulnar n.

Shoulder joint

Clavicle

Brachial plexus

Scalene muscles

ROOTS

TRUNKS

DIVISIONS

CORDS

Lung

Ribs

Musculo-cutaneous n.

Axillary n.

Radial n.

Ulnar n.

Median n.

Medial brachial cutaneous n.

Medial antebrachial cutaneous n.

First rib

C4

C5

C6

C7

C8

T1

N6: Brachial plexus

- The nerve roots of the lower cervical spinal cord split and merge several times before supplying the arm and hand.

- The brachial plexus lies over the first rib and behind the clavicle. It is intimately related to the subclavian/brachial artery and passes between the scalene muscles of the neck.

- The plexus is divided into roots, trunks, divisions, cords and terminal branches. By looking at the anatomical distribution of pain or dysfunction, it is possible to determine the location of a brachial plexus lesion.

- Brachial plexopathy can occur during delivery with or without shoulder dystocia, and from thoracic outlet syndrome.

NERVOUS SYSTEM

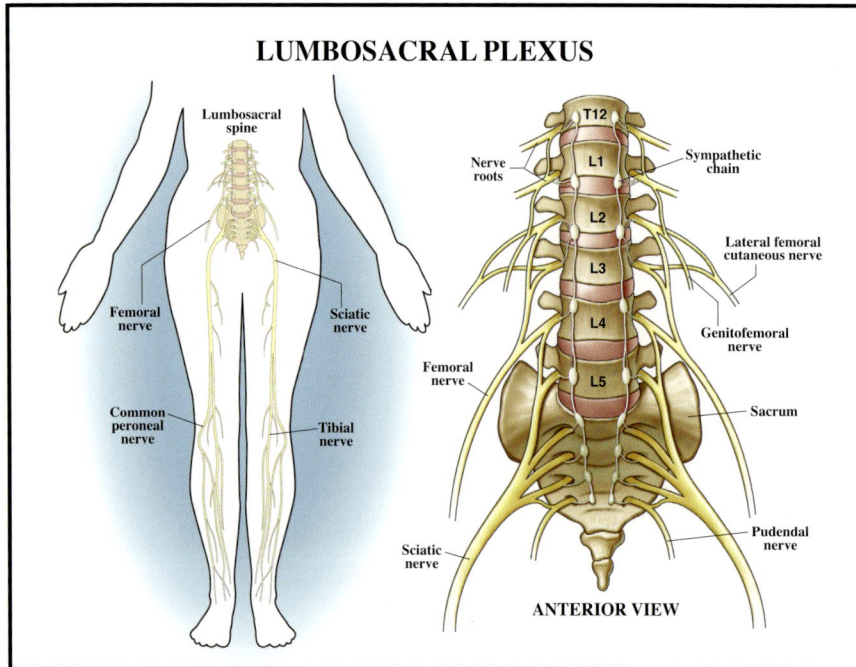

LUMBOSACRAL PLEXUS

Lumbosacral spine

Femoral nerve

Sciatic nerve

Common peroneal nerve

Tibial nerve

T12

Nerve roots

Sympathetic chain

L1

L2

Lateral femoral cutaneous nerve

L3

L4

Genitofemoral nerve

Femoral nerve

L5

Sacrum

Sciatic nerve

Pudendal nerve

ANTERIOR VIEW

N7: Lumbosacral plexus

- Analogous to the brachial plexus, the lumbosacral plexus is a series of nerve convergences and separations which ultimately combine into several large terminal nerves.

- Plexi form a protective mechanism in that if one nerve root is damaged, a particular muscle might be weakened, but function would not be completely lost.

- The terminal nerves in the legs generally follow the course of the deep vasculature.

- Terminal sensory nerves to the feet are particularly vulnerable to diabetes, resulting in peripheral diabetic neuropathy. This frequently contributes to foot infections and the need for amputation.

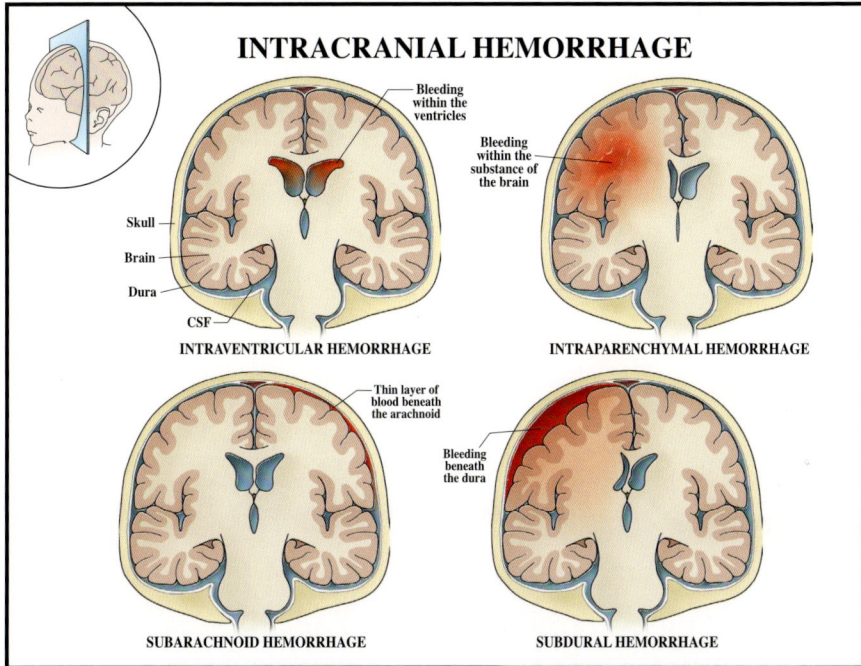

INTRACRANIAL HEMORRHAGE

Bleeding within the ventricles

Bleeding within the substance of the brain

Skull
Brain
Dura
CSF

INTRAVENTRICULAR HEMORRHAGE

INTRAPARENCHYMAL HEMORRHAGE

Thin layer of blood beneath the arachnoid

Bleeding beneath the dura

SUBARACHNOID HEMORRHAGE

SUBDURAL HEMORRHAGE

N8: Intracranial hemorrhage

- Intraventricular hemorrhage is bleeding within the cavities of the brain that normally hold clear cerebrospinal fluid (CSF). Such bleeding is frequently associated with pre-term delivery and can result in hydrocephalus and loss of brain tissue.

- Intraparenchymal bleeding is within the brain tissue itself and usually results from ruptured arteriovenous malformation (AVM), hemorrhage following ischemic infarction or hypertension.

- Subarachnoid hemorrhage usually results from a ruptured surface AVM or cerebral artery aneurysm.

- Subdural hemorrhage is the result of trauma leading to disruption of bridging veins between the dura and the brain.

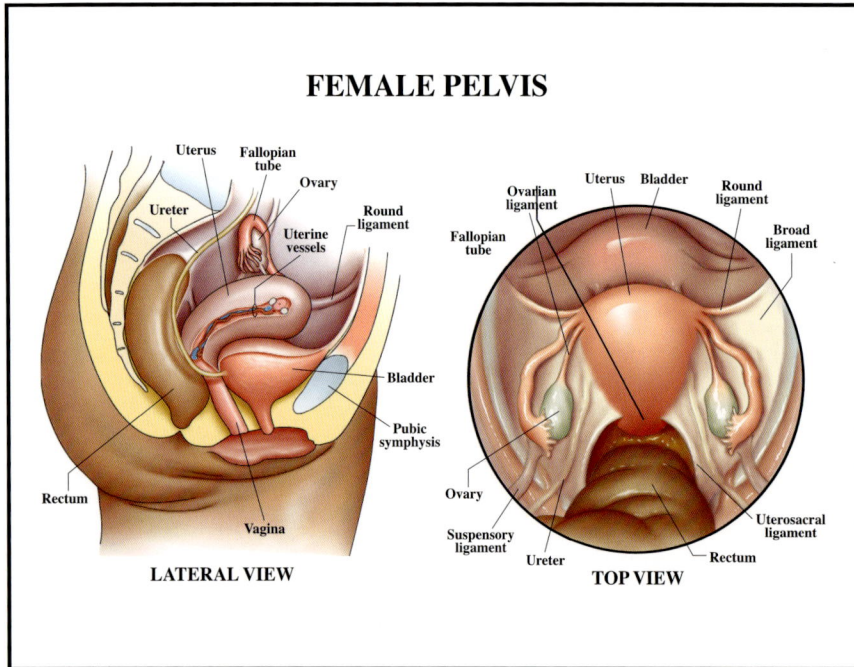

FEMALE PELVIS

Uterus
Fallopian tube
Ovary
Ureter
Uterine vessels
Round ligament
Bladder
Pubic symphysis
Rectum
Vagina

LATERAL VIEW

Ovarian ligament
Uterus
Bladder
Round ligament
Fallopian tube
Broad ligament
Ovary
Suspensory ligament
Ureter
Rectum
Uterosacral ligament

TOP VIEW

OG1: Female pelvis

- The female pelvis contains the bladder, uterus, vagina, and rectum. The tissue between the vaginal and rectal openings is a tight collection of tendons from the pelvic floor muscles, the perineum. The entire region is called the vulva.

- The non-pregnant uterus is about the size of a small pear. It is a hollow muscular organ, its neck enclosed by a thick circular muscle known as the cervix.

- Urine is excreted from the kidneys via the ureters, which transport it to the bladder. It is then carried to the outside by the relatively short urethra.

- The ovaries release ova (eggs) each month to the uterus via the fallopian tubes; ovarian hormones are absorbed into the bloodstream.

- The organs are held in the pelvis by a number of ligaments connecting them to the pelvic walls.

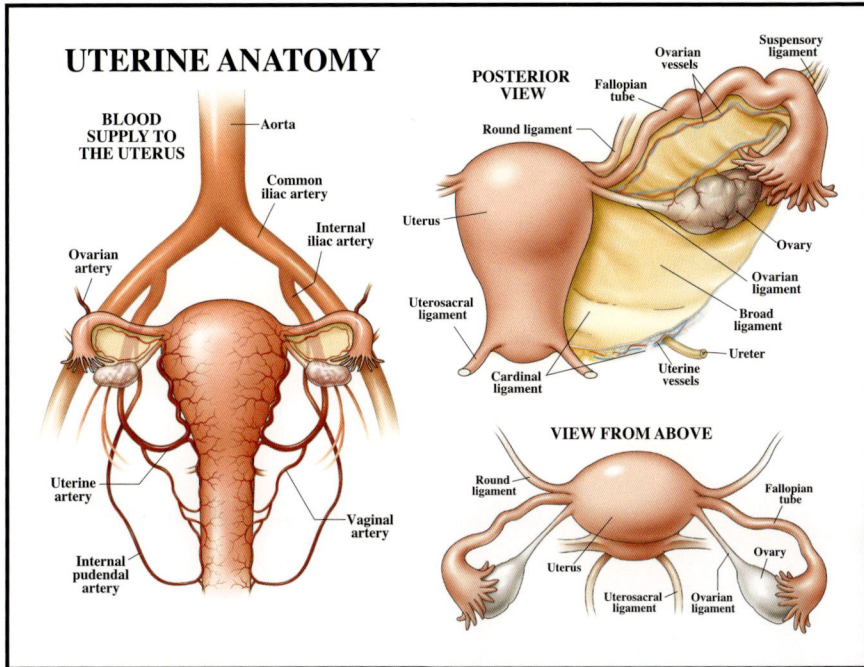

UTERINE ANATOMY

BLOOD SUPPLY TO THE UTERUS

Aorta

Common iliac artery

Internal iliac artery

Ovarian artery

Uterine artery

Internal pudendal artery

Vaginal artery

POSTERIOR VIEW

Round ligament

Uterus

Uterosacral ligament

Cardinal ligament

Ovarian vessels

Fallopian tube

Suspensory ligament

Ovary

Ovarian ligament

Broad ligament

Ureter

Uterine vessels

VIEW FROM ABOVE

Round ligament

Uterus

Uterosacral ligament

Ovarian ligament

Fallopian tube

Ovary

OG2: Uterine anatomy

- The uterus is small and contracted in its non-pregnant state, but grows to fill nearly the entire abdomen during pregnancy. Delivery is effected by repeated strong contractions pushing the fetus through the cervix and out the birth canal (vagina).

- The fallopian tubes carry the ova to the uterine cavity, where fertilized eggs implant and develop.

- The blood supply to the uterus is redundant and increases greatly during pregnancy.

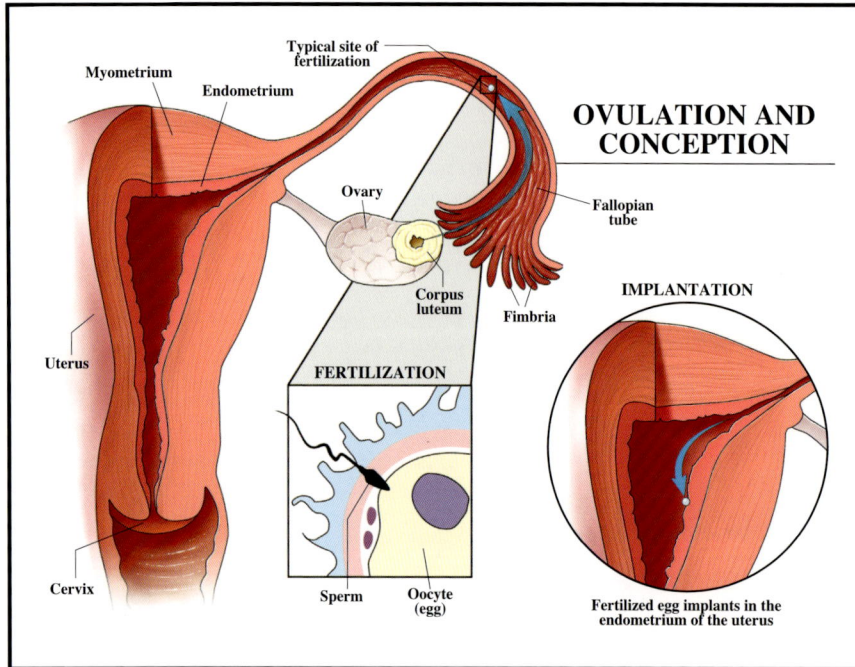

OVULATION AND CONCEPTION

Myometrium

Endometrium

Typical site of fertilization

Ovary

Corpus luteum

Fallopian tube

Fimbria

Uterus

FERTILIZATION

IMPLANTATION

Cervix

Sperm

Oocyte (egg)

Fertilized egg implants in the endometrium of the uterus

OG3: Ovulation and conception

- Fertilization usually occurs within the fallopian tube as the ovum is traveling toward the uterine cavity. If fertilization occurs, implantation occurs in about 7 days.

- The corpus luteum is a specialized area in the ovary which produces progesterone to support the early part of pregnancy; it forms at the development site of the recently released ovum.

- If the fallopian tube is scarred or blocked, as by pelvic inflammatory disease or structural deformity, ectopic pregnancy can result with likely tubal rupture, a potentially fatal condition for the mother.

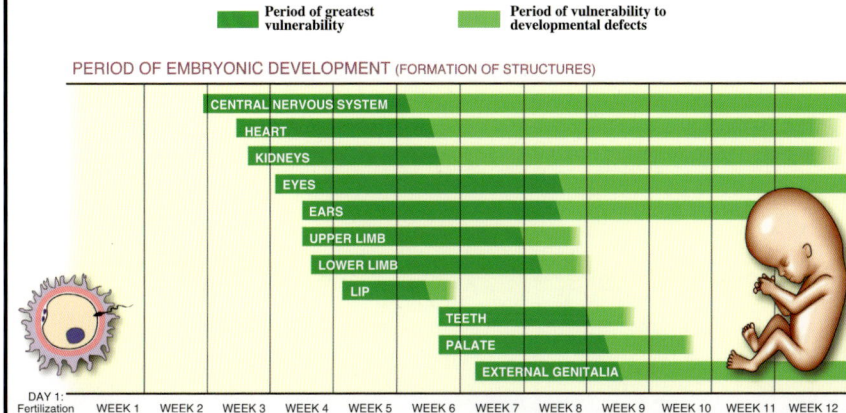

FETAL DEVELOPMENT I: First trimester

OG4: Fetal development I

- The first trimester is the time of development when the tissues specialize into specific organs. At the end of 12 weeks, all of the organs are present, albeit immature.

- Damage during the earliest weeks of gestation tends to be from infection or toxic exposure. Most of the time, this results in death and loss of the fetus.

- It is occasionally possible to date a damaging event. If, for example, there are heart, nervous system and limb deformations, it is likely that an insult occurred at around week 5.

- Fetuses delivered during the first trimester cannot survive.

FETAL DEVELOPMENT II: Second and third trimesters

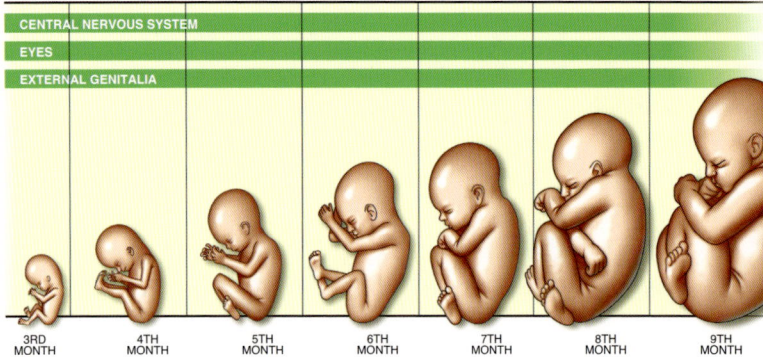

Legend:
- Period of greatest vulnerability
- Period of vulnerability to developmental defects

PERIOD OF FETAL GROWTH (MATURATION OF STRUCTURES)

CENTRAL NERVOUS SYSTEM

EYES

EXTERNAL GENITALIA

3RD MONTH · 4TH MONTH · 5TH MONTH · 6TH MONTH · 7TH MONTH · 8TH MONTH · 9TH MONTH

OG5: Fetal development II

- The second and third trimesters are devoted to growth and maturation. Fetuses may be viable outside the uterus from as early as 24 weeks, but brain and lung damage is common in fetuses delivered this early.

- The brain and lungs continue to mature even after delivery.

- A term pregnancy is considered to be 38-42 weeks long.

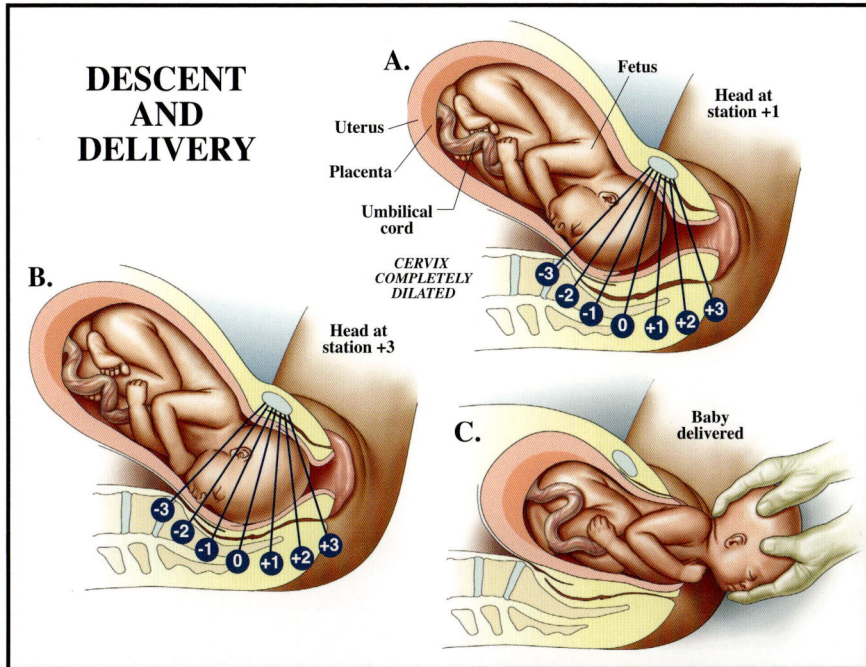

DESCENT AND DELIVERY

A.

Fetus

Uterus

Placenta

Umbilical cord

Head at station +1

CERVIX COMPLETELY DILATED

-3 -2 -1 0 +1 +2 +3

B.

Head at station +3

-3 -2 -1 0 +1 +2 +3

C.

Baby delivered

OG6: Descent and delivery

- In the first stage of labor, the circular muscle at the base of the uterus, the cervix, must thin (efface) and open (dilate) to a diameter of about 10 cm (4 in) to allow the fetal head to pass through.

- The cervix is opened by repeated uterine contractions pushing the presenting part of the fetus against the inside of the uterus, each time thinning and opening it more.

- The fetal presenting part is considered to be engaged (0 station) when the lowest portion is at the level of the ischial spines. Movement down the birth canal is measured by positive stations, some using a 0/+3 scale, and some using a 0/+5 scale.

- The second stage starts at full dilatation of the cervix and is completed when the fetus is delivered. The third stage is delivery of the placenta.

PREDICTING FETAL AGE AND WEIGHT

FUNDAL HEIGHT

Measurement from pubis to fundus in centimeters roughly equals the number of weeks of pregnancy

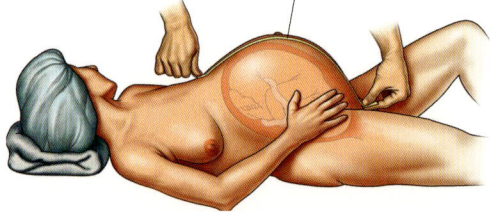

Symphysis-fundal measurement roughly equals the number of weeks of pregnancy

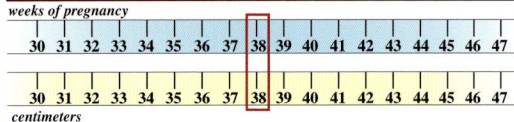

weeks of pregnancy

30 31 32 33 34 35 36 37 38 39 40 41 42 43 44 45 46 47

30 31 32 33 34 35 36 37 38 39 40 41 42 43 44 45 46 47

centimeters

ULTRASOUND

ABDOMINAL CIRCUMFERENCE

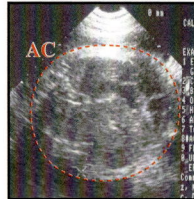

AC

BIPARIETAL DIAMETER

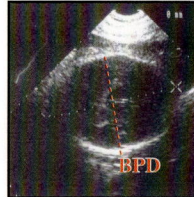

BPD

These values are plugged into a formula that determines fetal weight with 10% margin of error

OG7: Predicting fetal age/weight

- Fetal weight is notoriously difficult to predict.

- Fundal height is measured from the height of the top of the uterus to the pubic symphysis, and the height in cm is approximately equal to the number of weeks. The accuracy drops toward the end of the pregnancy as the fetus "drops" and enters the pelvis. Accuracy is also reduced in obese mothers.

- Ultrasound is very accurate in the first and second trimesters, but drops to about 65% as term approaches.

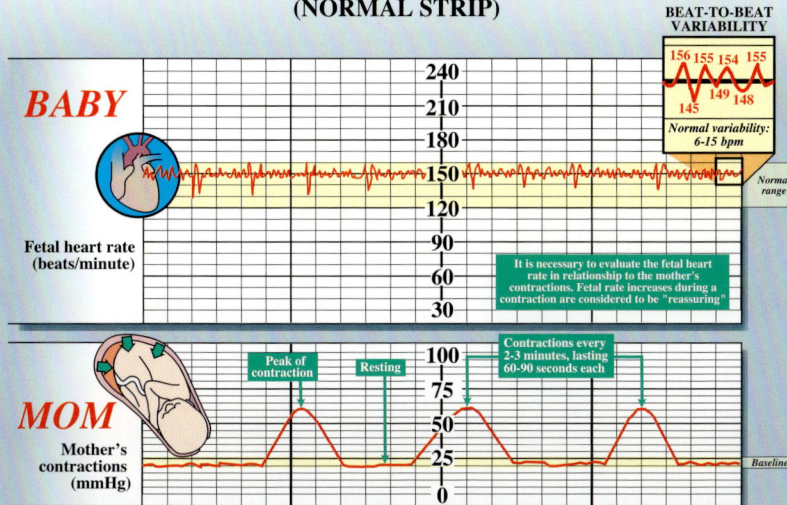

ELECTRONIC FETAL MONITORING
(NORMAL STRIP)

BABY

Fetal heart rate (beats/minute)

240 210 180 150 120 90 60 30

BEAT-TO-BEAT VARIABILITY

156 155 154 155
145 149 148

Normal variability:
6-15 bpm

Normal range

It is necessary to evaluate the fetal heart rate in relationship to the mother's contractions. Fetal rate increases during a contraction are considered to be "reassuring"

MOM

Mother's contractions (mmHg)

Peak of contraction

Resting

Contractions every 2-3 minutes, lasting 60-90 seconds each

100 75 50 25 0

Baseline

OG8: EFM (normal)

- This technology, used in approximately 87% of all labors in the U.S., tracks the fetal heart function against uterine contractions.

- The normal fetal heart rate is approximately 120-160 beats per minute (bpm), although normal individual fetuses might be higher or lower than this range.

- Beat-to-beat variability is literally the changes of the fetal heart rate from beat to beat (short-term variability), and within 3-5 minute periods (long-term variability). Beat-to-beat variability decreases or disappears for 20-30 minute time periods as the fetus sleeps, but is present in most normal labors and represents the health of the fetal brainstem.

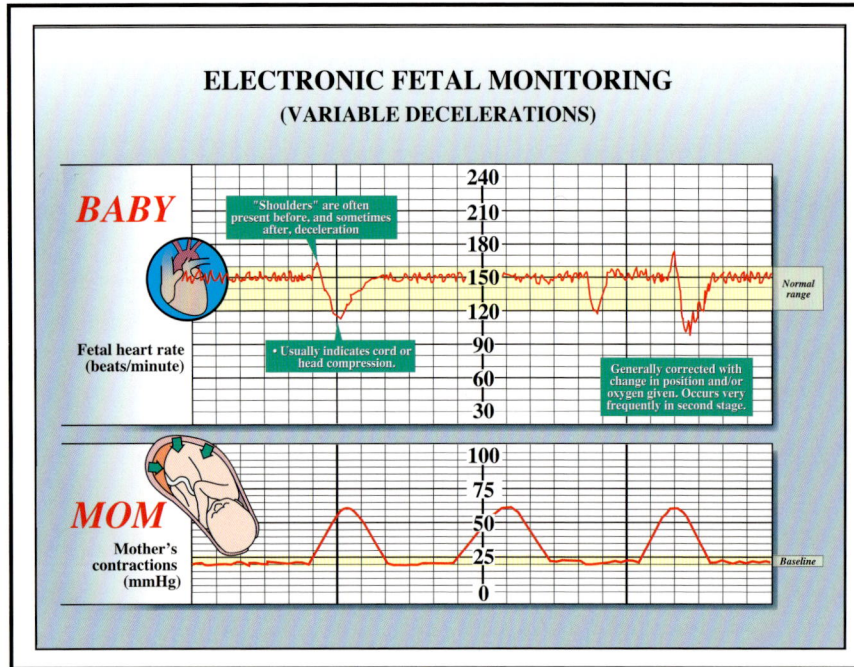

ELECTRONIC FETAL MONITORING
(VARIABLE DECELERATIONS)

BABY

"Shoulders" are often present before, and sometimes after, deceleration

240
210
180
150 — *Normal range*
120
90
60
30

Fetal heart rate (beats/minute)

• Usually indicates cord or head compression.

Generally corrected with change in position and/or oxygen given. Occurs very frequently in second stage.

MOM

Mother's contractions (mmHg)

100
75
50
25
0

Baseline

OG9: EFM (Variable decelerations)

• Variable decelerations can occur at any time during or between contractions, and are usually characterized by "shoulders" before and sometimes after the deceleration.

• The decelerations are usually "V" or "U"-shaped and return to baseline within two minutes or less.

• Variable decelerations are due to head or cord compression and are "treated" by changing the mother's position and applying oxygen.

• Unless very deep (<60 bpm) for extended periods (>2 minutes or more), they are considered benign.

• Variable decelerations occur during the second stage of most labors, as the fetal head moves down the narrow vaginal canal and is compressed by a combination of the uterine contractions and the narrow vagina.

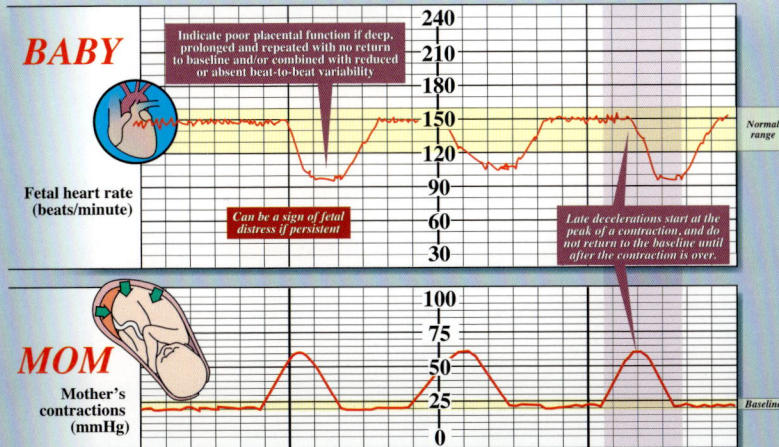

ELECTRONIC FETAL MONITORING
(LATE DECELERATIONS)

BABY

Indicate poor placental function if deep, prolonged and repeated with no return to baseline and/or combined with reduced or absent beat-to-beat variability

240
210
180
150
120
90
60
30

Normal range

Fetal heart rate (beats/minute)

Can be a sign of fetal distress if persistent

Late decelerations start at the peak of a contraction, and do not return to the baseline until after the contraction is over.

MOM

Mother's contractions (mmHg)

100
75
50
25
0

Baseline

OG10: EFM (Late decelerations)

• Late decelerations start at or after the peak of a contraction and are considered to be a sign of uteroplacental insufficiency.

• The depth of late decelerations is probably not as significant as their presence. If frequent, they can be a sign of fetal distress and an indication for prompt delivery.

• If late decelerations are accompanied by loss of beat-to-beat variability, it is generally considered an indication for urgent or emergent delivery, either by cesarean section or operative delivery (forceps or vacuum extraction), depending upon the state of the labor.

• The vast majority of fetuses with nonreassuring fetal heart tracings are completely normal.

APGAR SCORING SYSTEM

SCORE:	0	1	2		1 Min.	5 Min.	10 Min.
Heart Rate	Absent	<100	>100		2	2	2
Respiratory Effort	Absent	Slow, irregular	Good; crying		1	2	2
Muscle Tone	Floppy	Moderate	Good		1	1	1
Reflex Irritability	Unresponsive	Grimace	Strong cry		1	1	1
Color	Pale, bluish overall	Body pink; extremities dark blue	Completely pink			1	1

TOTAL SCORE: 5   7   7

EVALUATION OF TOTAL SCORE:
7 - 10:   Good
4 - 6:   Moderately depressed
0 - 3:   Severely depressed

OG11: Apgar scoring system

• This simple scoring system indicates how well the fetus fared through labor and delivery. The maximum score is 10 and the minimum is 0, with up to 2 points being awarded for each of 5 measures of health.

• Apgar scoring is done at 1 minute and 5 minutes of age, sometimes by nursing staff, sometimes by physicians. If the 5 minute score is low, a 10, 15 or 20 minute score may be recorded, either until the baby is stable or moved from the delivery room for specialized care.

• While there is some correlation between low 5 minute Apgar scores and neurological outcome, only a very small percentage of children with low 5 minute scores sustain brain damage. The longer the score remains low, the higher the correlation with neurological damage.

FORCEPS AND VACUUM DELIVERIES

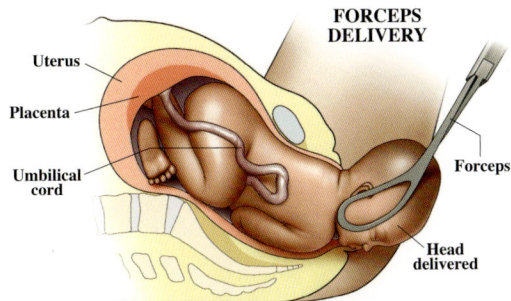

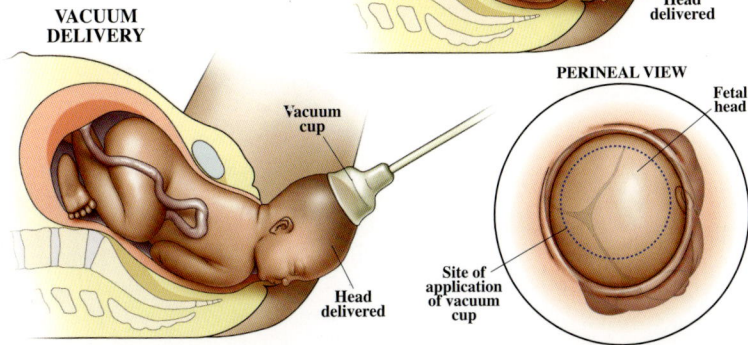

Uterus
Placenta
Umbilical cord

FORCEPS DELIVERY
Forceps
Head delivered

VACUUM DELIVERY
Vacuum cup
Head delivered

PERINEAL VIEW
Fetal head
Site of application of vacuum cup

OG12: Forceps/vacuum delivery

- Operative delivery is a vaginal delivery with the assistance of forceps or vacuum extraction.

- Forceps are of several types and the selection is determined by evaluation of the pelvis and presenting part, and the physician's training and preferences.

- The vacuum cap is a simple device which is placed on the vertex using a small hand-held vacuum pump; some physicians leave the vacuum on between contractions and some release the vacuum. The caps are set to pop off if the vacuum pressure is too high.

- Both procedures work with the uterine contractions, gently assisting and guiding the fetus out during contractions, and resting between them. Neither is used between contractions.

OBSTETRICS/GYNECOLOGY

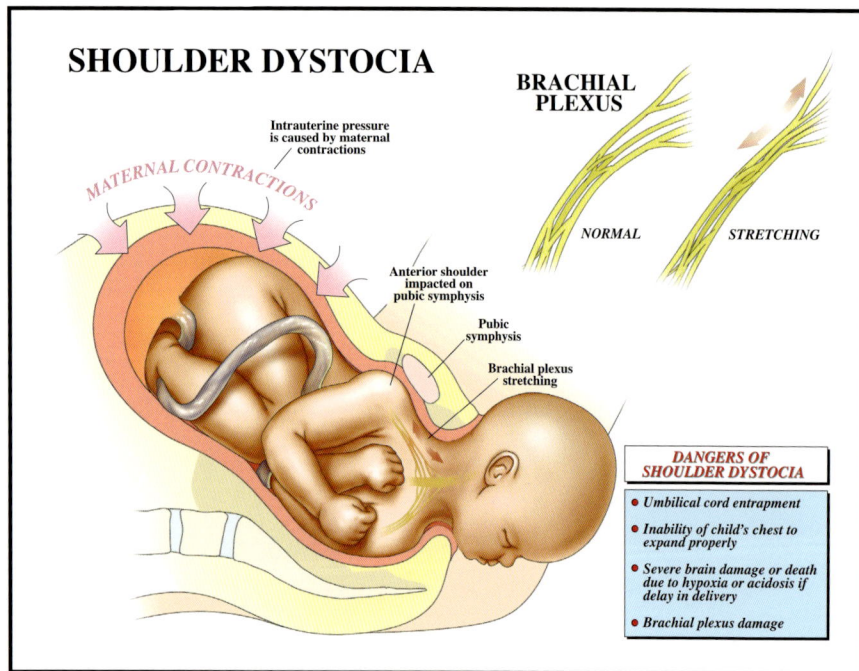

SHOULDER DYSTOCIA

Intrauterine pressure is caused by maternal contractions

MATERNAL CONTRACTIONS

BRACHIAL PLEXUS

NORMAL   STRETCHING

Anterior shoulder impacted on pubic symphysis

Pubic symphysis

Brachial plexus stretching

DANGERS OF SHOULDER DYSTOCIA
- Umbilical cord entrapment
- Inability of child's chest to expand properly
- Severe brain damage or death due to hypoxia or acidosis if delay in delivery
- Brachial plexus damage

OG13: Shoulder dystocia

- When shoulder dystocia occurs, the upper fetal shoulder is impacted on the pubic symphysis (more rarely, the lower shoulder on the sacral promontory or non-flexible coccyx), preventing delivery of the baby. This can be a potentially catastrophic event, since the fetal thorax is still within the pelvis and cannot properly expand for breathing.

- While the rate of shoulder dystocia is higher with gestational diabetic women and macrosomic fetuses (>4500 grams at birth), approximately 80% of shoulder dystocia occurs with average-sized fetuses.

- If a brachial plexus palsy (brachioplexopathy) occurs, it usually affects the portions of the brachial plexus which control the shoulder and elbow. Spontaneous recovery is the rule rather than the exception.

- Most fetuses are successfully delivered with a combination of McRobert's maneuver and suprapubic pressure, with or without episiotomy.

72

DELIVERY FORCES

Normal systolic blood pressure (strength of left ventricle contactions) is ~110 mmHg

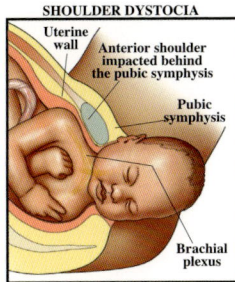

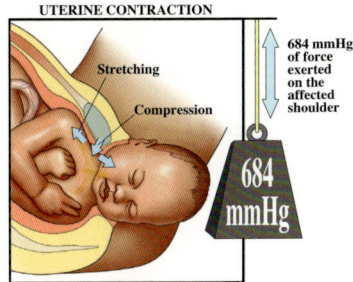

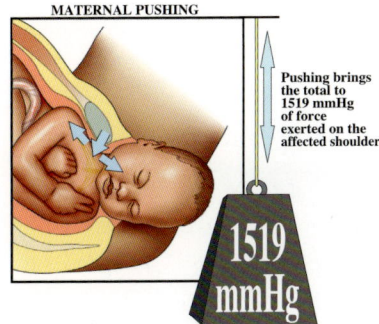

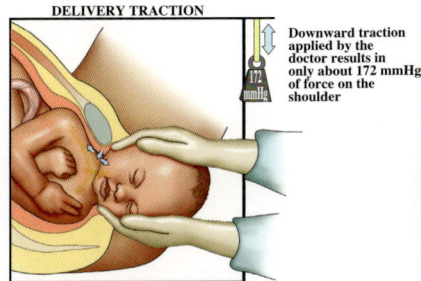

SHOULDER DYSTOCIA
Uterine wall
Anterior shoulder impacted behind the pubic symphysis
Pubic symphysis
Brachial plexus

UTERINE CONTRACTION
Stretching
Compression
684 mmHg of force exerted on the affected shoulder
684 mmHg

MATERNAL PUSHING
Pushing brings the total to 1519 mmHg of force exerted on the affected shoulder
1519 mmHg

DELIVERY TRACTION
Downward traction applied by the doctor results in only about 172 mmHg of force on the shoulder
172 mmHg

OG14: Delivery forces

- The uterus itself generates tremendous pressure during contractions. If the average systolic blood pressure (pressure generated by the left ventricle during a contraction) is 120 mmHg, the uterus alone generates more than 5 times that amount. When the accessory muscles are used -- the diaphragm and abdominal muscles -- the pressure increases to more than 10 times that amount.

- The pull generated by a physician during downward traction for shoulder dystocia is slightly higher than systolic pressure and contributes only a tiny amount to the total amount of generated pressures.

- Uterine and abdominal pressures are good evidence that if shoulder dystocia is the cause of brachial plexus palsy, it is most likely from the intrinsic pressures of the uterus and body wall, not the caregiver.

73

OBSTETRICS/GYNECOLOGY

MECONIUM:
Timing the Insult

Stress can trigger the release of meconium (intestinal contents) into the amniotic fluid

1 For the first hours after release, meconium floats in clumps

Uterus
Placenta
Meconium

Membrane (containing amniotic fluid)

2 After 6-12 hours, the meconium resembles "pea soup." This can last for several days.

3 After several days, the meconium starts to clear. The baby, amniotic sac and surrounding fluid are stained.

4 After several weeks, the fluid is clear. The baby's skin also clears, but stains are left on the nail beds, mucous membranes and amniotic sac.

Umbilical cord

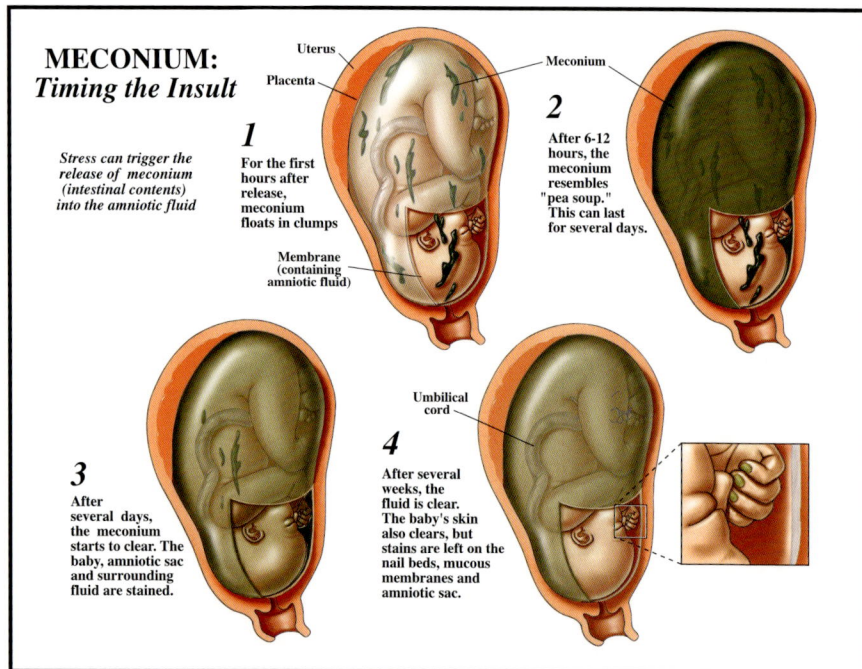

OG15: Meconium

• Meconium is the dark, tarry contents of the fetal gastrointestinal tract, and is composed of fluids, swallowed cells and debris. Meconium passage at labor and delivery is common, occurring in up to 20% of deliveries; this is not a sign of fetal distress unless accompanied by other signs.

• Meconium release prior to labor is considered a sign of fetal distress. When meconium is freshly released, it is clumpy and floats in the amniotic fluid. Within hours, it distributes throughout the sac and is known as "pea-soup" meconium. Macrophages in the fetal skin, mucus membranes and the amniotic sac phago-cytize particulate matter, giving a green cast to the tissues.

• After several days, the fluid is greenish-brown but clear, and the fetus and membranes are stained. After several weeks, the fetal skin and fluid will clear, but the amniotic membranes, fetal mucus membranes and nailbeds remain stained.

74

AMERICAN COLLEGE OF OBSTETRICS AND GYNECOLOGY
AMERICAN ACADEMY OF PEDIATRICS
International Cerebral Palsy Task Force
ACOG Cerebral Palsy Task Force
2003

In order to make plausible link between perinatal asphyxia and neurologic deficit in an individual patient, __all four__ of the following criteria must be present:

☐ **Metabolic acidosis in umbilical artery blood at delivery** (*umbilical artery pH <7, base deficit >12*)

☐ **Spastic quadriplegia or dyskinetic cerebral palsy**

☐ **Early neurological signs in infants >34 weeks of gestation** (*coma, seizures, floppiness, loss of reflexes*)

☐ **Exclusion of all other identifiable causes** (*trauma, bleeding disorders, infection, genetic disorders, etc.*)

OG16: ACOG/AAP Definition

- The American College of Obstetrics and Gynecology and the American Academy of Pediatrics have recommended adoption of a definition of term intrapartum asphyxia developed by an international task force in 2003.

- The definition requires that all four parameters must be met in order to diagnose asphyxiation in a term fetus. These include metabolic acidosis (represented by umbilical artery pH of <7.0 and base excess of <-12/base deficit of >12); specific types of cerebral palsy correlated with the kinds of asphyxial damage seen in term fetal brains; early seizures and other neurological signs, and exclusion of all other causes.

- These criteria do not apply to pre-term infants, since their immature systems result in different signs.

OBSTETRICS/GYNECOLOGY

PLACENTAL ABRUPTION

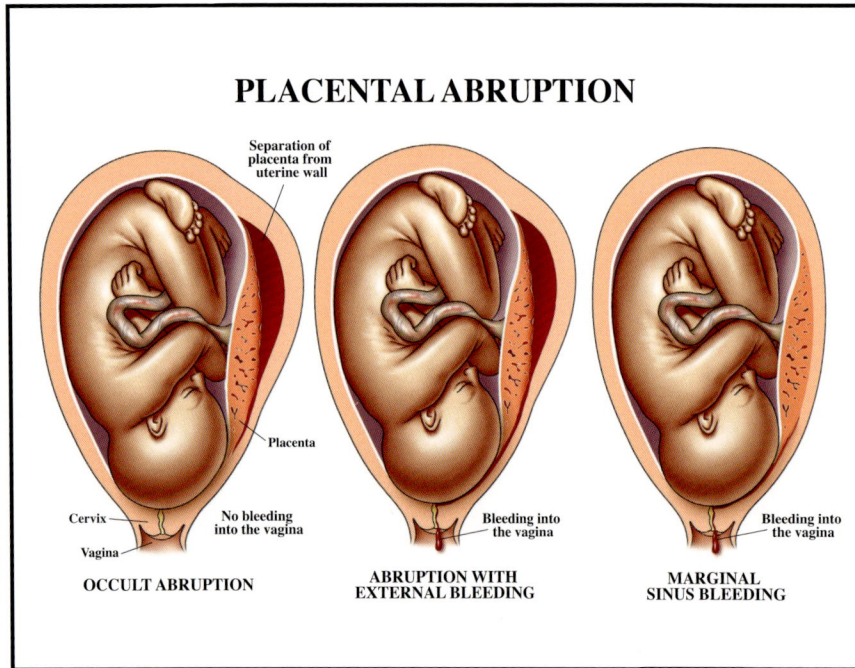

Separation of placenta from uterine wall

Placenta

Cervix

Vagina

No bleeding into the vagina

OCCULT ABRUPTION

Bleeding into the vagina

ABRUPTION WITH EXTERNAL BLEEDING

Bleeding into the vagina

MARGINAL SINUS BLEEDING

OG17: Placental abruption

• Abruption occurs when the placenta separates from the uterine wall. Most of the time, this is accompanied by vaginal bleeding as the blood travels between the membranes and the uterus. Sometimes, however, there is no bleeding because edges of the placenta remain sealed.

• Symptoms frequently include a hypertonic uterus with severe abdominal pain. If the abruption is large (more than about 50%), the fetus may not survive.

• Causes include abdominal trauma, maternal hypertension, maternal cocaine or heroine use and cigarette use. In many cases, a direct cause is never determined.

• Sometimes there appears to be significant vaginal bleeding, but all fetal and maternal signs are normal. This marginal sinus separation is benign.

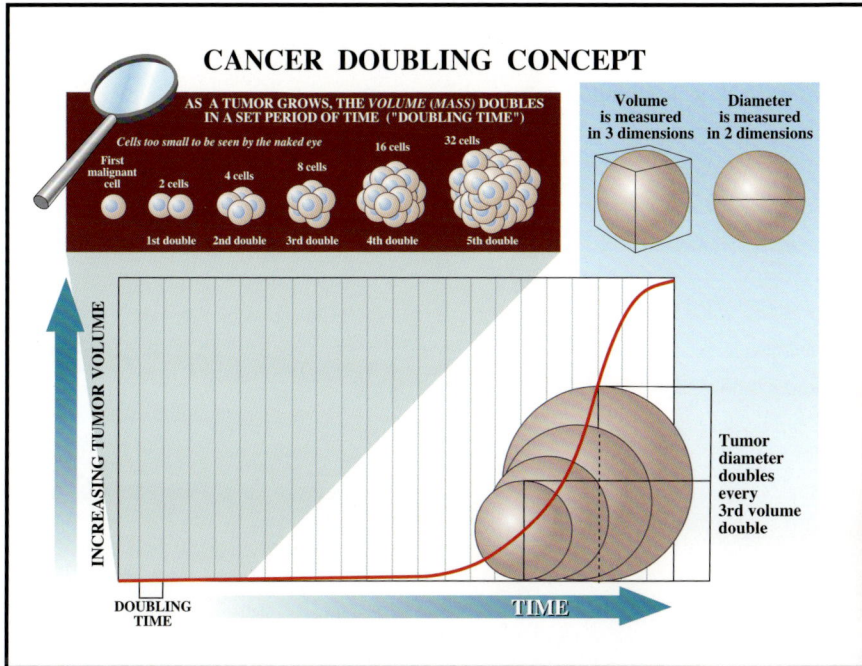

CANCER DOUBLING CONCEPT

AS A TUMOR GROWS, THE *VOLUME (MASS)* DOUBLES IN A SET PERIOD OF TIME ("DOUBLING TIME")

Cells too small to be seen by the naked eye

First malignant cell

2 cells — 1st double
4 cells — 2nd double
8 cells — 3rd double
16 cells — 4th double
32 cells — 5th double

Volume is measured in 3 dimensions

Diameter is measured in 2 dimensions

INCREASING TUMOR VOLUME

DOUBLING TIME

TIME

Tumor diameter doubles every 3rd volume double

ON1: Cancer doubling concept

• Most solid tumors tend to double in volume at approximately the same rate throughout the majority of their life spans, with possible slow-down very early and/or at the end.

• There is a small body of literature published in the 1960s-80s in which the growth of these tumors was measured over time; means and ranges were calculated for the more common tumor types. These curves tend to follow a bell-shaped curve, adding to the likelihood of their reliability.

• Generally, the more rapid the doubling time, the more aggressive the tumor. The average breast cancer doubles about every 125 days, for example, but very aggressive tumors have been described as doubling as rapidly as every 2-3 days; less aggressive at every 300 days or more.

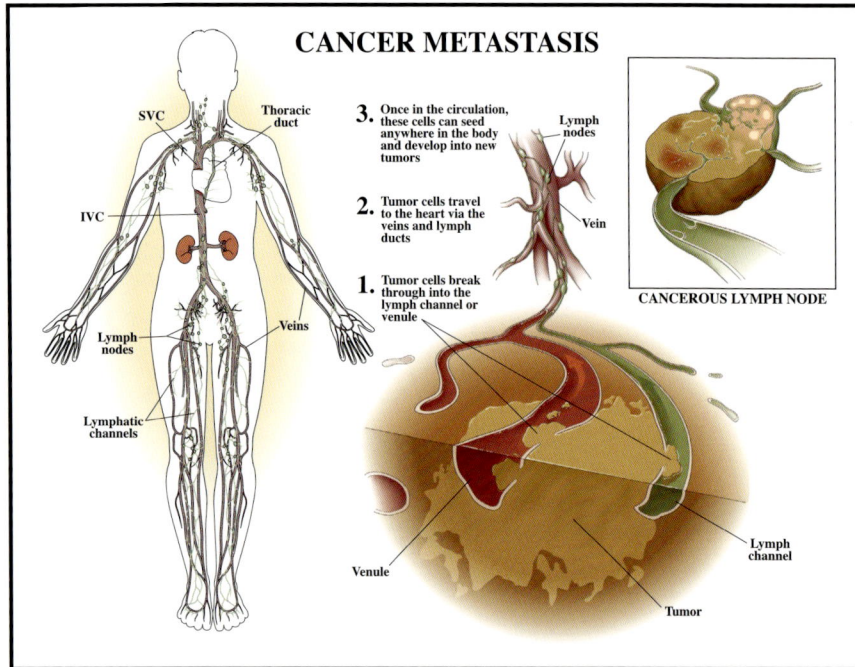

CANCER METASTASIS

SVC

Thoracic duct

IVC

3. Once in the circulation, these cells can seed anywhere in the body and develop into new tumors

2. Tumor cells travel to the heart via the veins and lymph ducts

1. Tumor cells break through into the lymph channel or venule

Lymph nodes

Vein

Lymph nodes

Veins

Lymphatic channels

CANCEROUS LYMPH NODE

Venule

Lymph channel

Tumor

ON2: Cancer metastasis

• Cancer tends to spread by two mechanisms: infiltration, in which the tumor pushes against and enters contiguous tissue; and metastasis, when cancer cells enter lymphatic channels and/or small blood vessels and eventually travel to distant locations and organs in the body.

• Certain tumors have a predilection for specific sites. Colon cancer frequently spreads to the liver, and breast cancer to the brain and spine.

• When there is lymphatic spread, the local and regional lymph nodes are the first line of defense; when the nodes fill up with dividing tumor cells, the cells then break free and travel toward the heart for distribution throughout the body.

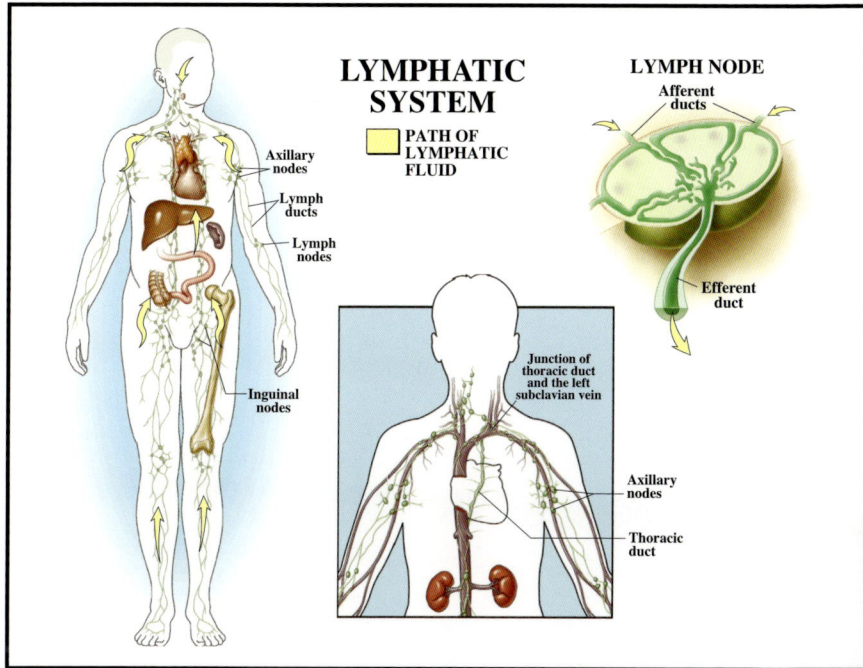

LYMPHATIC SYSTEM

□ **PATH OF LYMPHATIC FLUID**

Axillary nodes

Lymph ducts

Lymph nodes

Inguinal nodes

LYMPH NODE

Afferent ducts

Efferent duct

Junction of thoracic duct and the left subclavian vein

Axillary nodes

Thoracic duct

ON3: Lymphatic system

• The lymphatic system is part of the body's complex immune system. Lymph is a clear fluid which travels in tiny channels along the major vessels toward the heart. There are two large thoracic ducts which collect lymph and drain into the circulatory system at the subclavian veins.

• The lymph nodes are pea-sized collections of tissue located throughout the lymphatic system in chains and clusters. They serve as filters to catch bacteria, cancer cells or other foreign material.

• Regional lymph nodes filter lymph from a particular body region. Axillary nodes, for example, collect lymph draining from the breast and are a protective mechanism against early breast cancer.

• Sentinel nodes are the first node metastatic cells reach, and there are new imaging techniques which can help determine if metastases have occurred.

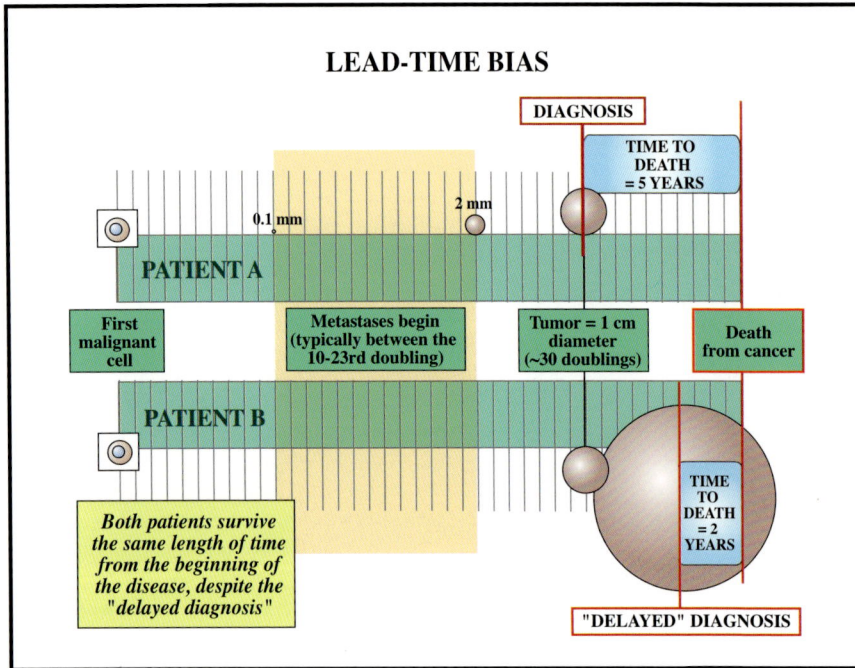

LEAD-TIME BIAS

DIAGNOSIS

TIME TO DEATH = 5 YEARS

0.1 mm

2 mm

PATIENT A

First malignant cell

Metastases begin (typically between the 10-23rd doubling)

Tumor = 1 cm diameter (~30 doublings)

Death from cancer

PATIENT B

Both patients survive the same length of time from the beginning of the disease, despite the "delayed diagnosis"

TIME TO DEATH = 2 YEARS

"DELAYED" DIAGNOSIS

ON4: Lead-time bias

- A common assumption is that a cancer may not have been diagnosed early enough to make a difference, with the assumption being that early diagnosis is always better.

- In reality, cancers are present for very long periods of time before they are diagnosable, and many have metastasized prior to the period in which they can be detected.

- The life span of more than 60% of cancer patients is essentially predetermined by the characteristics of the cancer itself. While a patient diagnosed earlier may live "longer" than one diagnosed later, both patients actually survive about the same length of time from the first cancer cell.

BREAST ANATOMY AND LYMPHATICS

ANTERIOR VIEW

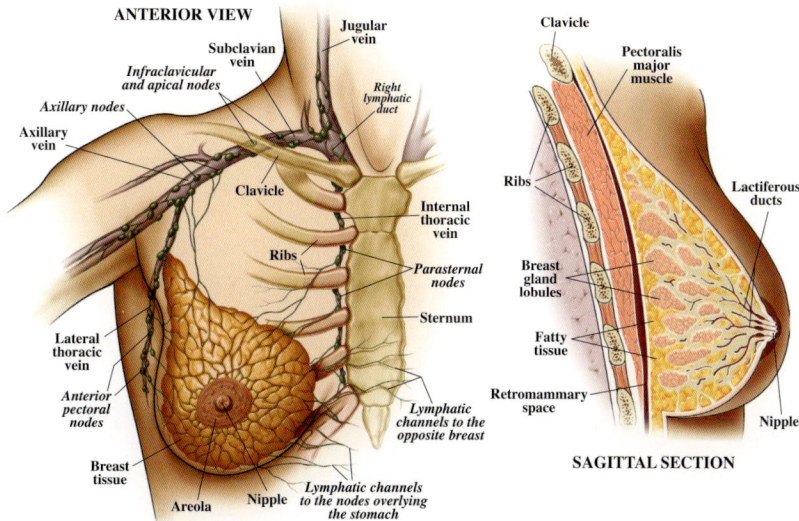

Labels (anterior view): Subclavian vein, Jugular vein, Infraclavicular and apical nodes, Axillary nodes, Right lymphatic duct, Axillary vein, Clavicle, Internal thoracic vein, Ribs, Parasternal nodes, Sternum, Lateral thoracic vein, Anterior pectoral nodes, Breast tissue, Areola, Nipple, Lymphatic channels to the opposite breast, Lymphatic channels to the nodes overlying the stomach

SAGITTAL SECTION

Labels (sagittal section): Clavicle, Pectoralis major muscle, Ribs, Lactiferous ducts, Breast gland lobules, Fatty tissue, Retromammary space, Nipple

ON5: Breast anatomy/lymphatics

- The breast in a premenopausal woman is composed of glandular tissue, fat, connective tissue and ducts; the axillary tail of breast tissue is tucked upward in the axilla. The breast lies on the pectoralis muscles of the thorax.

- The breast is divided by irregular fibrous septa which prevent masses from migrating from one area of the breast to another; malignant tumors may eventually erode through these septa.

- Lymphatic channels travel throughout the breast, with all but the most medial portions draining to the axillary lymph nodes. The "sentinel" node—the first node to receive drainage from the breast—can be determined with testing and evaluated for metastatic spread.

- Post-menopausal women have little glandular tissue since most of it has been replaced by fat.

COLORECTAL CANCER

TNM	Modified Astler-Coller	Dukes'
T1N0M0	MAC A	Dukes' A
T2N0M0	MAC B1	
T3N0M0	MAC B2	Dukes' B
T4N0M0	MAC B3	
T2N1M0	MAC C1	Dukes' C
T2N2M0		
T3N1M0		
T2N2M0		
T3N1M0	MAC C2	
T3N2M0		
T4N1M0	MAC C3	
T4N2M0		
T(any)N(any)M1		Dukes' D

M (Distant metastases)
M0 = no distant metastasis
M1 = distant metastases present

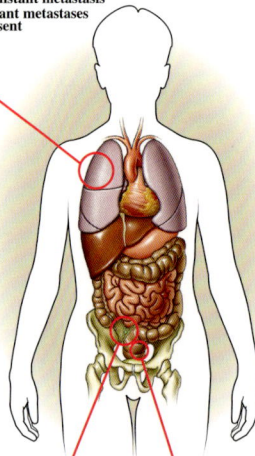

N (Regional nodes)
N0 = no evidence of tumor
N1 = 1-3 positive nodes
N2 = 4 or more positive nodes
N3 = central nodes positive

T (Primary tumor)
T0 = no evidence of tumor
T1 = invades submucosa
T2 = invades muscularis
T3 = through muscularis
T4 = into surrounding tissues

ON6: Colorectal cancer staging

- Tumors are "staged" in order to help design a treatment plan. The TNM system is common to all solid tumor classification systems, with T showing the size of the primary tumor, N indicating the presence of affected regional lymph nodes, and M showing the presence or absence of distant metastases.

- The staging is then classified into categories which have treatment and prognostic significance. In the case of colorectal cancer, two of the more commonly used staging systems are the Astler-Coller and Dukes' methods.

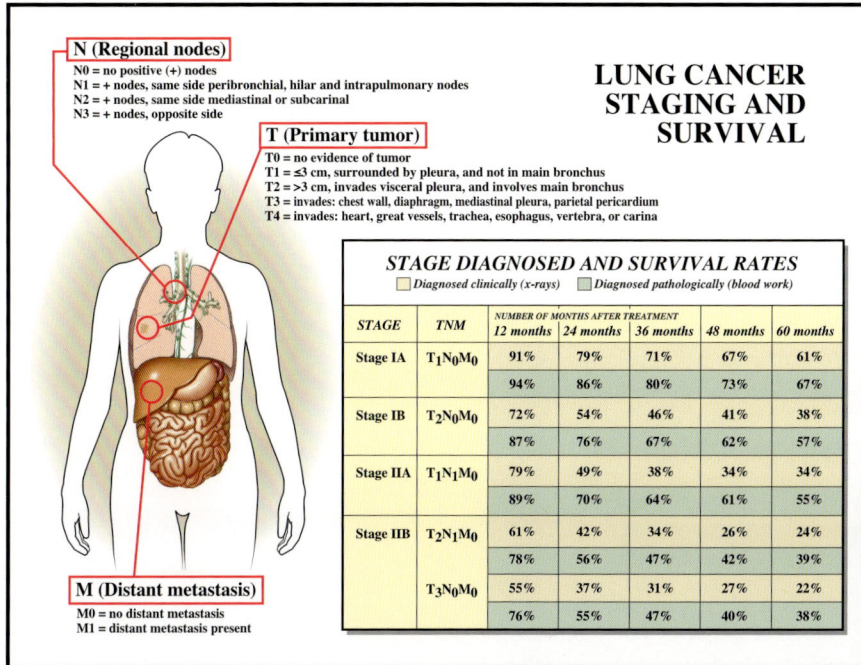

N (Regional nodes)
N0 = no positive (+) nodes
N1 = + nodes, same side peribronchial, hilar and intrapulmonary nodes
N2 = + nodes, same side mediastinal or subcarinal
N3 = + nodes, opposite side

T (Primary tumor)
T0 = no evidence of tumor
T1 = ≤3 cm, surrounded by pleura, and not in main bronchus
T2 = >3 cm, invades visceral pleura, and involves main bronchus
T3 = invades: chest wall, diaphragm, mediastinal pleura, parietal pericardium
T4 = invades: heart, great vessels, trachea, esophagus, vertebra, or carina

LUNG CANCER STAGING AND SURVIVAL

M (Distant metastasis)
M0 = no distant metastasis
M1 = distant metastasis present

STAGE DIAGNOSED AND SURVIVAL RATES

Diagnosed clinically (x-rays)   Diagnosed pathologically (blood work)

STAGE	TNM	NUMBER OF MONTHS AFTER TREATMENT				
		12 months	24 months	36 months	48 months	60 months
Stage IA	$T_1N_0M_0$	91%	79%	71%	67%	61%
		94%	86%	80%	73%	67%
Stage IB	$T_2N_0M_0$	72%	54%	46%	41%	38%
		87%	76%	67%	62%	57%
Stage IIA	$T_1N_1M_0$	79%	49%	38%	34%	34%
		89%	70%	64%	61%	55%
Stage IIB	$T_2N_1M_0$	61%	42%	34%	26%	24%
		78%	56%	47%	42%	39%
	$T_3N_0M_0$	55%	37%	31%	27%	22%
		76%	55%	47%	40%	38%

ON7: Lung cancer staging

- Tumors are "staged" in order to help design a treatment plan. The TNM system is common to all solid tumor classification systems, with T showing the size of the gross tumor, N indicating the presence of affected regional lymph nodes, and M showing the presence or absence of distant metastases.

- The staging is then classified into categories which have treatment and prognostic significance. One of the key issues in the staging of lung cancer is whether the cancer has spread to the other side of the chest (contralateral nodes), a generally poor prognostic sign.

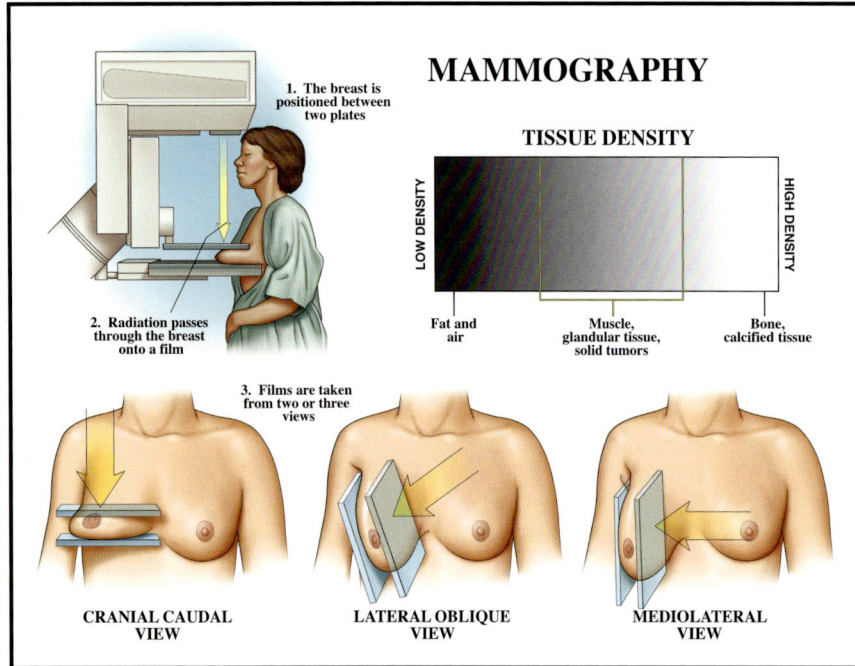

MAMMOGRAPHY

1. The breast is positioned between two plates

2. Radiation passes through the breast onto a film

3. Films are taken from two or three views

TISSUE DENSITY

LOW DENSITY

HIGH DENSITY

Fat and air

Muscle, glandular tissue, solid tumors

Bone, calcified tissue

CRANIAL CAUDAL VIEW

LATERAL OBLIQUE VIEW

MEDIOLATERAL VIEW

ON8: Mammography

- Mammography is an imaging technique which allows visualization of breast tissue. Fat, glandular tissue and ductal tissue have characteristic densities and patterns.

- Solid tumors are generally easier to see in older patients with larger amounts of fat within the breasts, but can be difficult to see in patients with "dense" breasts (young women and women with fibrocystic breasts).

- Approximately 75% of breast cancers can be seen on mammography, with the patient's breast characteristics being the largest determining factor.

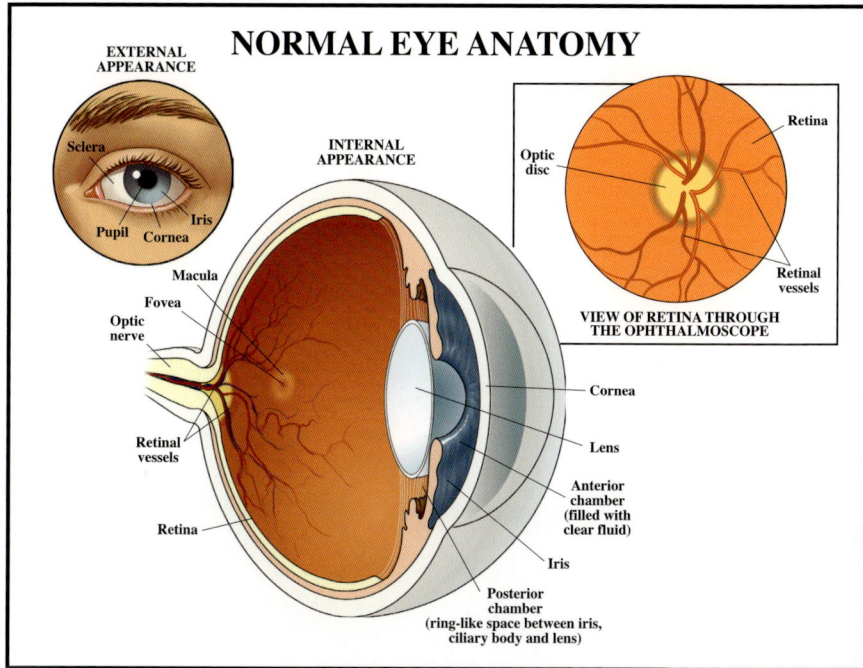

NORMAL EYE ANATOMY

EXTERNAL APPEARANCE

Sclera
Pupil
Iris
Cornea

INTERNAL APPEARANCE

Macula
Fovea
Optic nerve
Retinal vessels
Retina

Cornea
Lens
Anterior chamber (filled with clear fluid)
Iris
Posterior chamber (ring-like space between iris, ciliary body and lens)

Optic disc
Retina
Retinal vessels

VIEW OF RETINA THROUGH THE OPHTHALMOSCOPE

OPH1: Normal eye anatomy

• The external eye has upper and lower lids which close over the globe to protect it. The sclera is the white of the eye, the colored portion is the iris, and the black opening in the middle of the iris is a hole known as the pupil. This is the only window in the body through which the nervous system can be seen directly.

• The anterior transparent media consists of the cornea, anterior chamber and lens; the posterior elements of the globe are covered with specialized nerve tissue, the retina.

• The optic nerve enters the eye posteriorly along with its own blood supply; this area is known as the optic disc. The macular area is where visual acuity is greatest.

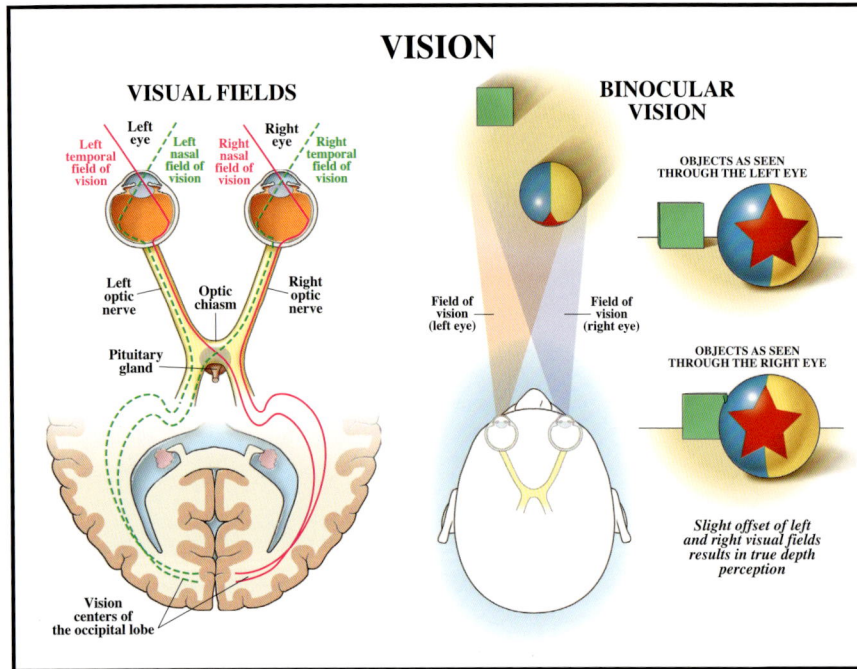

VISION

VISUAL FIELDS

Left temporal field of vision
Left eye
Left nasal field of vision
Right nasal field of vision
Right eye
Right temporal field of vision

Left optic nerve
Optic chiasm
Right optic nerve

Pituitary gland

Vision centers of the occipital lobe

BINOCULAR VISION

OBJECTS AS SEEN THROUGH THE LEFT EYE

Field of vision (left eye)
Field of vision (right eye)

OBJECTS AS SEEN THROUGH THE RIGHT EYE

Slight offset of left and right visual fields results in true depth perception

OPH2: Vision

- The visual system is composed of specialized nerve fibers originating in the retina. They join to form the optic nerve (cranial nerve II) just behind the globe. The medial portions of the optic nerve cross over each other at the optic chiasm in front of the pituitary gland. As the nerve fibers travel posteriorly within the brain, they form the optic tracts terminating in the occipital lobe of the brain. By identifying visual field defect patterns, it is possible to determine an anatomical location of the source of visual loss.

- The visual fields of normally-aligned eyes overlap. Each eye sees objects from a slightly different angle, and the brain fuses these views. This binocular vision allows us to perceive depth and spatial relationships.

CATARACT SURGERY

① Incision made in conjunctiva

Conjunctival incision

Limbus

② Opening made in anterior capsule of lens

Anterior capsule incision

Limbal incision

③ Nucleus of lens brought out through limbal incision

Limbal incision lengthened

④ Cortex of lens removed

Limbal incision partially closed

Cavitron

Cortex

⑤ Intraocular lens placed

Lens implant

Conjunctival and limbal incisions sutured

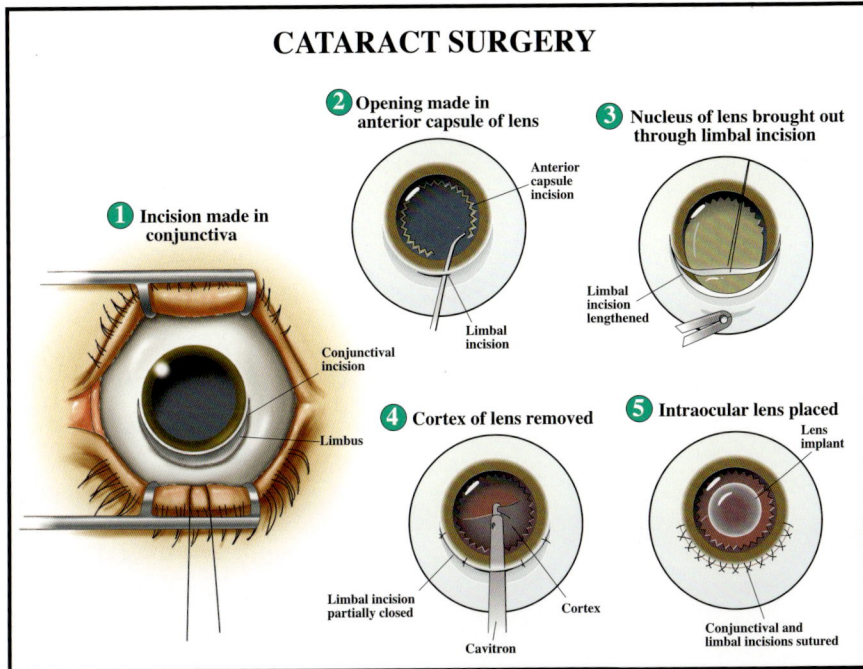

OPH3: Cataract surgery

• The lens opacifies with age or after trauma. When vision is sufficiently affected, the cataract can be surgically removed and replaced with an artificial intraocular lens.

• The cornea is lifted from an incision in the blue-grey line surrounding the iris, and the anterior surface of the lens is opened. The nucleus of the lens is removed, leaving the posterior capsule of the lens in position.

• The intraocular lens is then placed within the capsule and fixed into position. Laser treatments are often needed post-operatively to clear the posterior capsule.

• This procedure is one of the safest and most common surgical procedures performed today, with a very low rate of complications.

MYOPIA AND HYPEROPIA

MYOPIA (Nearsightedness)	NORMAL EYE	HYPEROPIA (Farsightedness)
Image forms in front of retina	Image forms on retina	Image forms behind retina

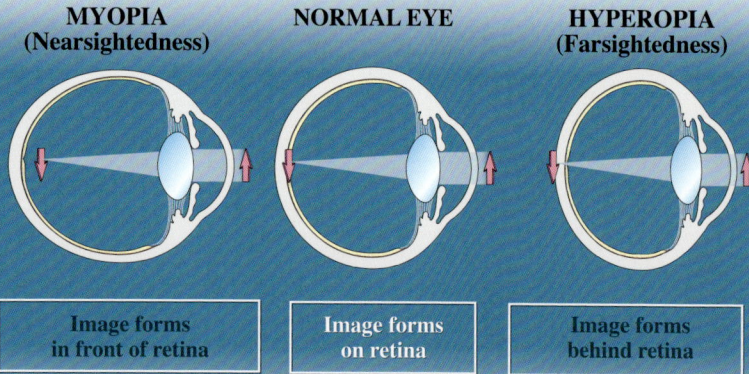

OPH4: Myopia/hyperopia

- Visual acuity depends upon the clarity of the transparent media at the front of the eye (cornea, anterior chamber fluid and lens), the clarity of the viscous vitreous in the globe, and the distance of the macula from the lens.

- In the normal eye, light rays focus in a point at the macula, the area on the retina which allows the greatest acuity. In nearsightedness, the eyeball is elongated, or the lens is defective and the focal point lies in front of the macula. In farsightedness, the opposite condition exists. Both can be corrected with lenses, and sometimes the newer laser corneal surgery techniques can be used.

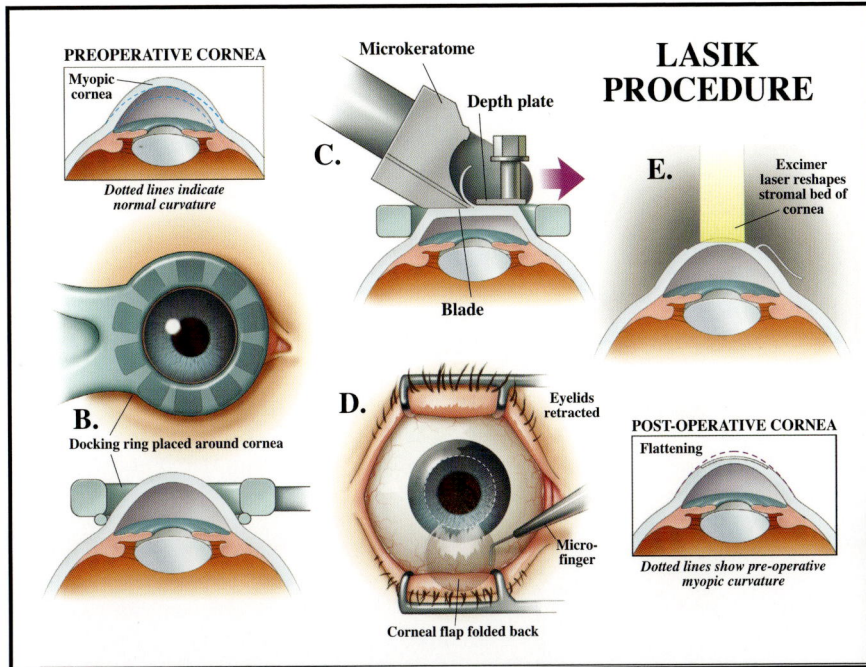

PREOPERATIVE CORNEA

Myopic cornea

Dotted lines indicate normal curvature

B.

Docking ring placed around cornea

Microkeratome

Depth plate

C.

Blade

D.

Eyelids retracted

Micro-finger

Corneal flap folded back

LASIK PROCEDURE

E.

Excimer laser reshapes stromal bed of cornea

POST-OPERATIVE CORNEA

Flattening

Dotted lines show pre-operative myopic curvature

OPH5: LASIK procedure

- LASIK (laser-assisted in situ keratomileusis) is a popular type of refractive surgery, or surgery performed to improve visual acuity.

- In LASIK, an incision is made to lift up a partial thickness of the cornea, using a very sharp, thin microtome.

- Once the flap is formed, the stroma of the cornea is sculpted with the laser, under computer control. Many of the newer LASIK systems can also accommodate for any eye movement during surgery, using a tracking program.

- This procedure has a very high success rate with relatively few complications.

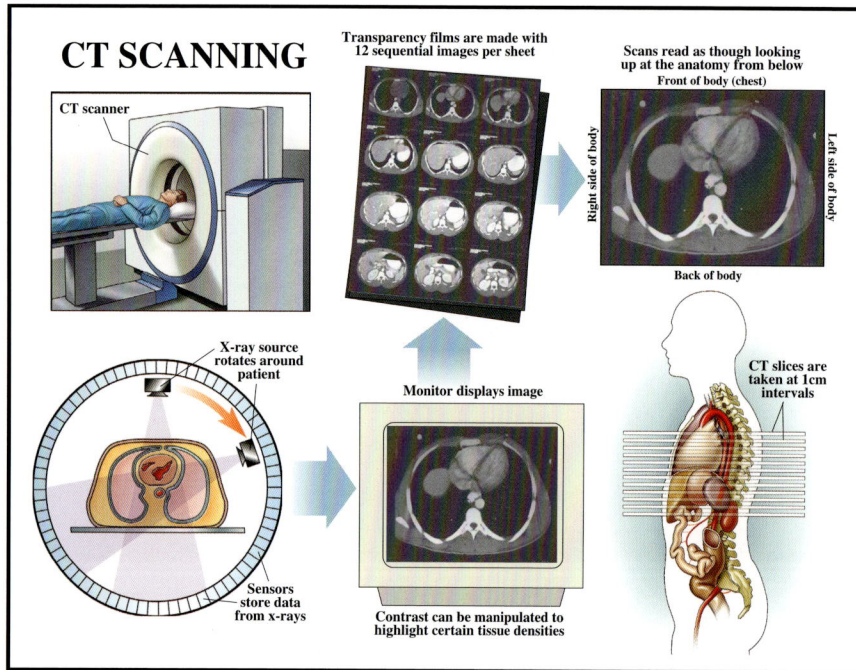

CT SCANNING

CT scanner

X-ray source rotates around patient

Sensors store data from x-rays

Transparency films are made with 12 sequential images per sheet

Monitor displays image

Contrast can be manipulated to highlight certain tissue densities

Scans read as though looking up at the anatomy from below
Front of body (chest)

Right side of body

Left side of body

Back of body

CT slices are taken at 1cm intervals

Rad 1: CT scanning

- CT, or computerized axial tomography, is a method of x-ray imaging in which the x-rays enter the body from 360 degrees, as the radiation source rotates around the patient.

- The computer calculates a series of slices, which are examined individually or can be reconstructed into a three-dimensional form. They are usually read directly from the monitor and are then printed on x-ray film.

- MRI scanning is similar in that it also generates a 360 degree scan, but it utilizes a magnetic field which makes the tissues vibrate at different frequencies. These signals are translated into gradations on the monitor, and can also be printed to x-ray film.

- Both CT and MRI images are generally presented in the form of contiguous "slices" of the region of interest.

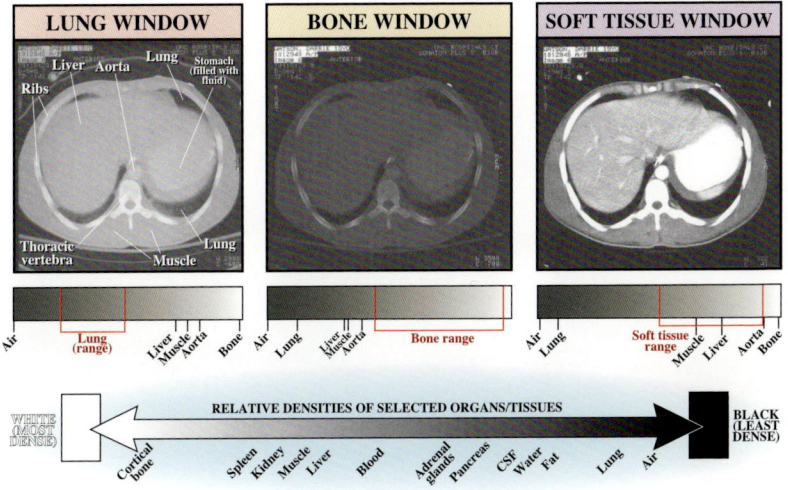

CT WINDOW ILLUSTRATIONS

LUNG WINDOW — **BONE WINDOW** — **SOFT TISSUE WINDOW**

RELATIVE DENSITIES OF SELECTED ORGANS/TISSUES

Rad2: CT windows

• The data collected by CT can be modified to show certain structures more clearly. Since this is usually at the expense of other tissues, the slices are often printed several times, using different "windows" or exposures.

• Lung windows are set to show the air-filled lungs clearly; lungs would normally be dark on regular radiographic imaging.

• Bone windows make all soft tissue a nearly uniform shade, but bone is very bright and clearly seen.

• Soft-tissue windows allow some distinction between different densities of soft tissue, making edema, tumors and other abnormalities more obvious.

• Other specialized settings are available for MRI.

RESPIRATORY

ANATOMY OF RESPIRATION

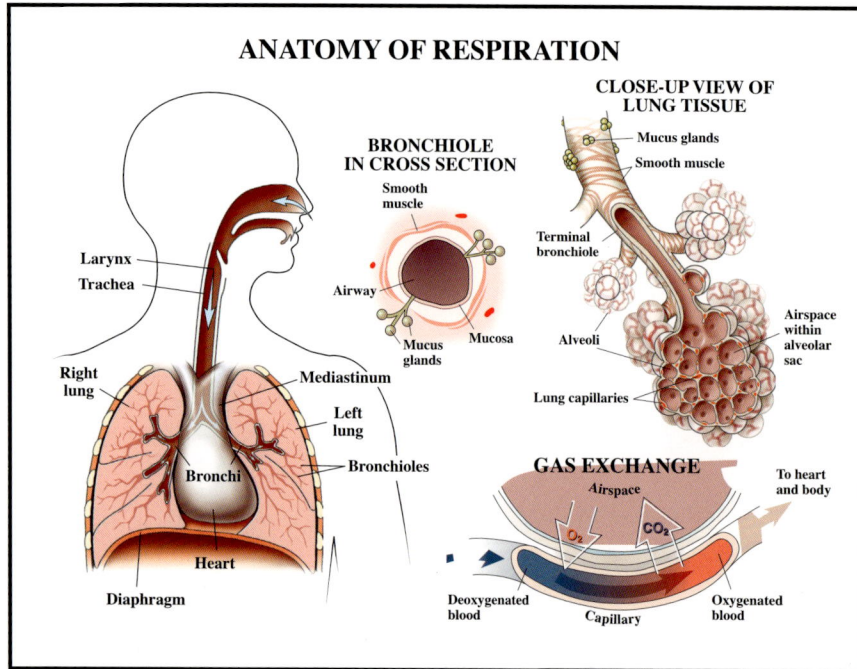

BRONCHIOLE IN CROSS SECTION

Smooth muscle

Airway

Mucus glands

Mucosa

CLOSE-UP VIEW OF LUNG TISSUE

Mucus glands

Smooth muscle

Terminal bronchiole

Alveoli

Lung capillaries

Airspace within alveolar sac

Larynx

Trachea

Right lung

Mediastinum

Left lung

Bronchioles

Bronchi

Heart

Diaphragm

GAS EXCHANGE

Airspace

O_2

CO_2

To heart and body

Deoxygenated blood

Capillary

Oxygenated blood

R1: Anatomy of respiration

- The lungs are composed of thin-walled alveoli whose sacs are covered by a mesh-work of capillaries. This is where oxygen and carbon dioxide are exchanged.

- The trachea carries air from the nose and mouth to the bronchi, which branch to each lung. These divide several times to become very small bronchioles, which directly supply the alveoli.

- The airways are lined with a ciliated mucosa which carries debris upward to the mouth on a layer of mucous, where it is swallowed. These mucosal membranes can swell in reaction to allergens, bacteria and viruses, leading to narrow airways and respiratory symptoms.

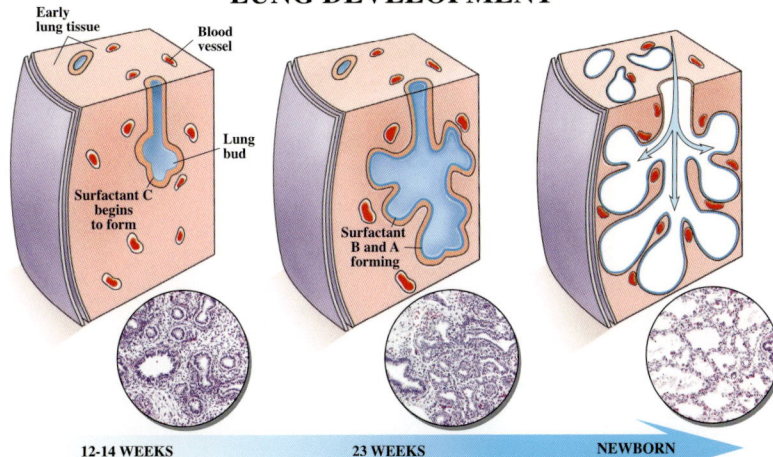

LUNG DEVELOPMENT

Early lung tissue

Blood vessel

Lung bud

Surfactant C begins to form

Surfactant B and A forming

12-14 WEEKS

- Lung buds grow into early lung tissue
- Thickened lung bud and blood vessel walls
- Surfactant C starts to form

23 WEEKS

- Lungs continue to grow into early lung tissue
- Surfactant B and A forming

NEWBORN

- Surfactant aids in the inflation of alveoli
- Alveolar and blood vessel walls only 1 cell thick allowing air exchange

R2: Lung development

- The main reason that preterm infants are considered high risk is because their lungs are immature.

- Lungs develop as the airways bud and branch into an anlage of mesenchymal cells. Since respiration requires oxygen and carbon dioxide to cross over two layers of tissue (alveolar wall and capillary wall), these relatively thick-walled airways in preterm babies permit little gas exchange. High-pressure ventilation is required to assist the infant, and this pressure frequently results in the development of chronic lung disease (bronchopulmonary dysplasia).

- In addition, there are too few alveoli present for efficient oxygen supply until 2-3 weeks prior to term. Lungs continue to grow and develop new alveoli for several years after birth.

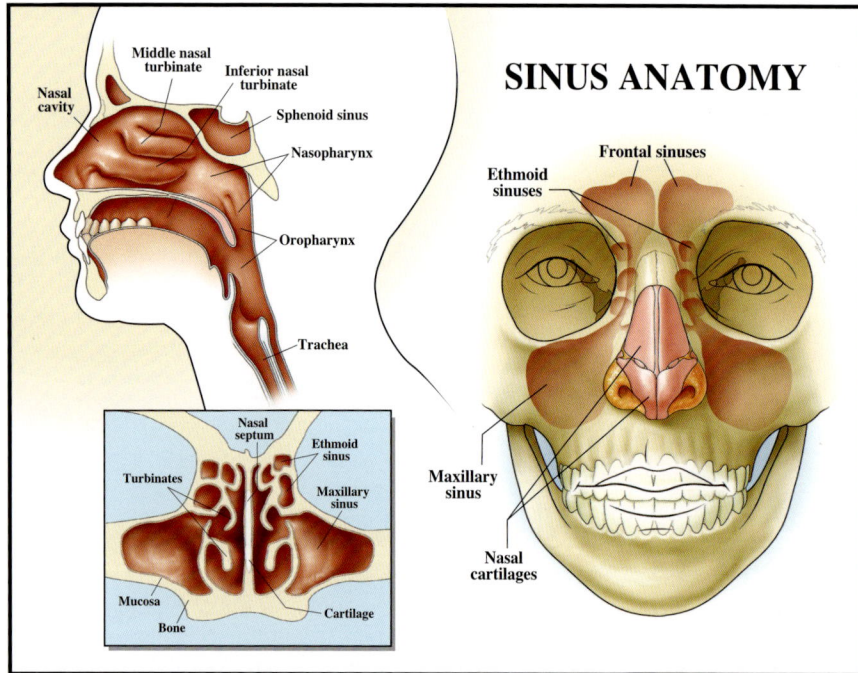

SINUS ANATOMY

Middle nasal turbinate
Inferior nasal turbinate
Nasal cavity
Sphenoid sinus
Nasopharynx
Oropharynx
Trachea

Frontal sinuses
Ethmoid sinuses
Maxillary sinus
Nasal cartilages

Nasal septum
Ethmoid sinus
Turbinates
Maxillary sinus
Mucosa
Bone
Cartilage

R3: Sinus anatomy

- Sinuses are hollow spaces within the facial bones. They are lined with a ciliated mucosa which has mucus glands. The sinuses are interconnected via a series of openings, allowing mucus to drain into the nose and pharynx.

- The sinuses help to warm inhaled air before it enters the lungs.

- Sinuses are prone to infection or reaction to allergens and react by mucosal swelling and overproduction of mucus. Chronic inflammation or infection can result in permanent thickening of the mucosa and reactive bone changes. Surgery is designed to facilitate drainage and relieve pressure; in some patients it must be repeated a large number of times.

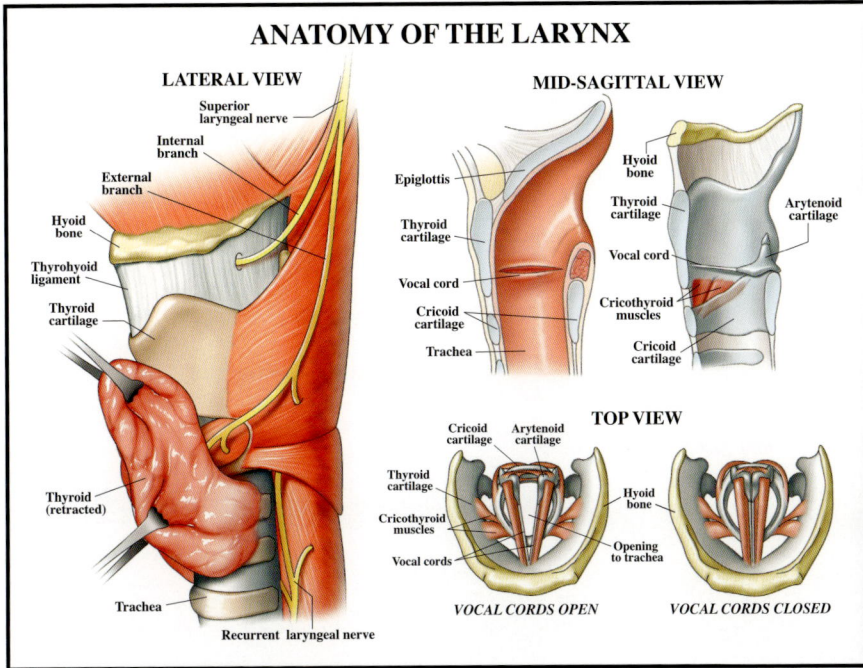

ANATOMY OF THE LARYNX

LATERAL VIEW

- Superior laryngeal nerve
- Internal branch
- External branch
- Hyoid bone
- Thyrohyoid ligament
- Thyroid cartilage
- Thyroid (retracted)
- Trachea
- Recurrent laryngeal nerve

MID-SAGITTAL VIEW

- Epiglottis
- Thyroid cartilage
- Vocal cord
- Cricoid cartilage
- Trachea
- Hyoid bone
- Thyroid cartilage
- Arytenoid cartilage
- Vocal cord
- Cricothyroid muscles
- Cricoid cartilage

TOP VIEW

- Cricoid cartilage
- Arytenoid cartilage
- Thyroid cartilage
- Cricothyroid muscles
- Vocal cords
- Hyoid bone
- Opening to trachea

VOCAL CORDS OPEN *VOCAL CORDS CLOSED*

R4: Anatomy of the larynx

• The larynx is composed of a number of cartilaginous structures, muscles and ligaments which maintain the patency of the airway and hold the vocal cords under tension during speech.

• The large thyroid cartilage, which lies beneath the thyroid gland, is connected to the hyoid bone by a strong ligament (thyrohyoid ligament), and the epiglottis arises from its internal surface. All internal structures with the exception of the vocal cords are covered by a pink mucosal lining.

• The small cartilages to which the vocal cords are attached are moved by tiny muscles under the control of the recurrent, superior and inferior laryngeal nerves. These muscles make small adjustments in the opening between the cords, allowing different pitches of sound to be created.

ENDOTRACHEAL INTUBATION

LARYNGOSCOPIC VIEW

Trachea

Esophagus

Lung

Laryngoscope

ET Tube

Epiglottis

Trachea

Esophagus

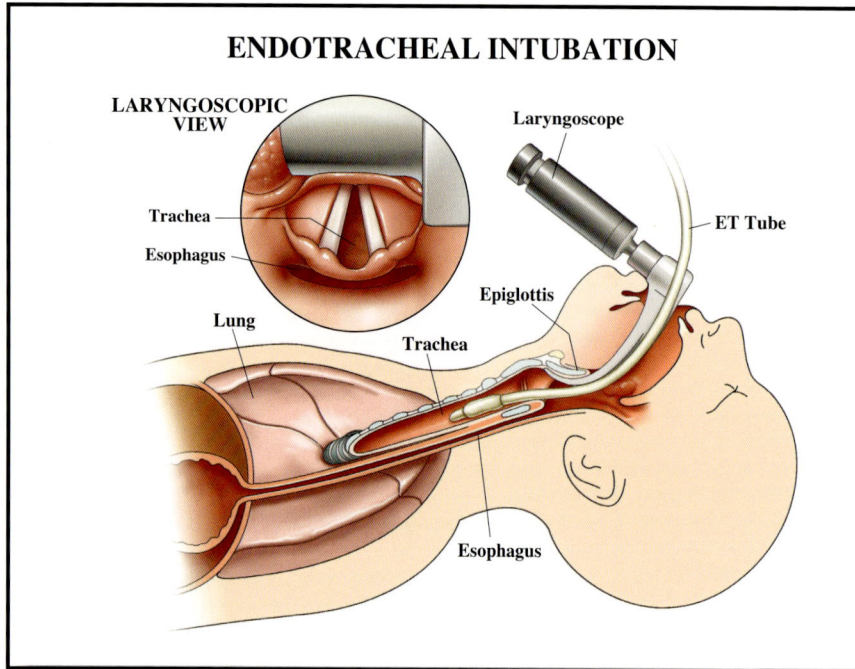

R5: Endotracheal intubation

• Intubation is required when a patient has difficulty breathing and needs ventilatory assistance. A hollow tube is inserted into the trachea and held in place by a small inflated balloon. If intubation is required for more than a few weeks, a tracheostomy is used to replace it.

• Most endotracheal intubations are done using a laryngoscope, which holds the tongue and epiglottis out of the way while the health care provider inserts the ETT (endotracheal tube).

• Following ETT placement, the provider listens for bilateral breath sounds, watches for the chest to rise, and usually orders a portable chest x-ray to check ETT placement.

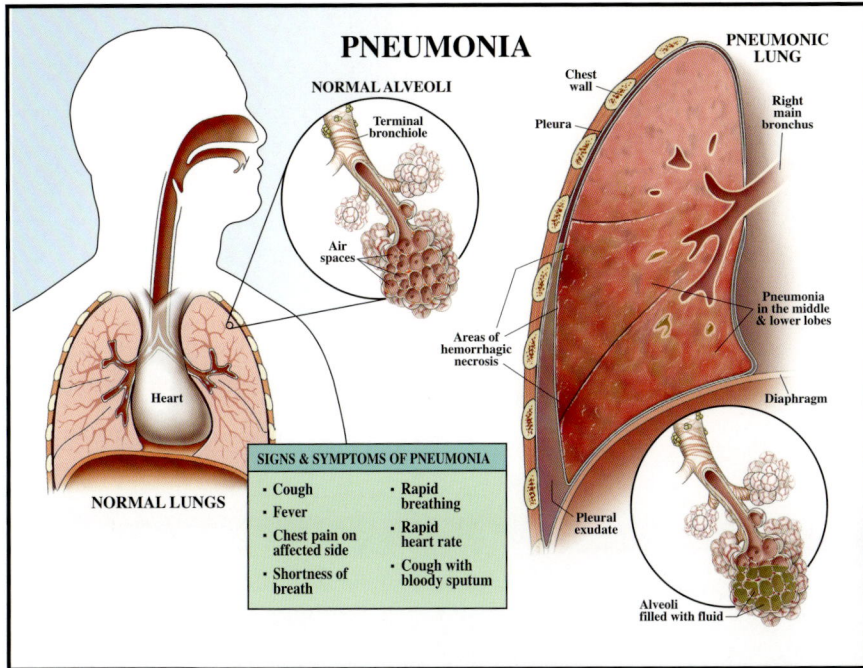

PNEUMONIA

NORMAL ALVEOLI

PNEUMONIC LUNG

Terminal bronchiole

Chest wall

Pleura

Right main bronchus

Air spaces

Areas of hemorrhagic necrosis

Pneumonia in the middle & lower lobes

Diaphragm

Heart

NORMAL LUNGS

Pleural exudate

Alveoli filled with fluid

SIGNS & SYMPTOMS OF PNEUMONIA

- Cough
- Fever
- Chest pain on affected side
- Shortness of breath
- Rapid breathing
- Rapid heart rate
- Cough with bloody sputum

R6: Pneumonia

- Pneumonia is a potentially serious condition in which the alveoli thicken and fill with fluid and pus.

- The lungs themselves get a "meaty" texture and may have small abscesses or areas of infection.

- Pneumonia may be caused by bacteria, viruses, fungi or parasites, and its form is slightly different for each cause, although the symptoms are very much the same. Similar findings can also result from chemicals or toxins.

- Pneumonia is often accompanied by fever, cough, air hanger with rapid breathing, changes on exam and changes on chest x-ray.

UROGENITAL

THE URINARY SYSTEM

Kidneys

Ureter

Bladder

Renal artery

Renal vein

Ureter

Pyramid

Cortex

Medulla

Renal pelvis

Ureter

SECTION

Afferent arteriole

Glomerulus

Efferent arteriole

Proximal tubule

Distal tubule

Loop of Henle

NEPHRON (FUNCTIONAL UNIT OF THE KIDNEY)

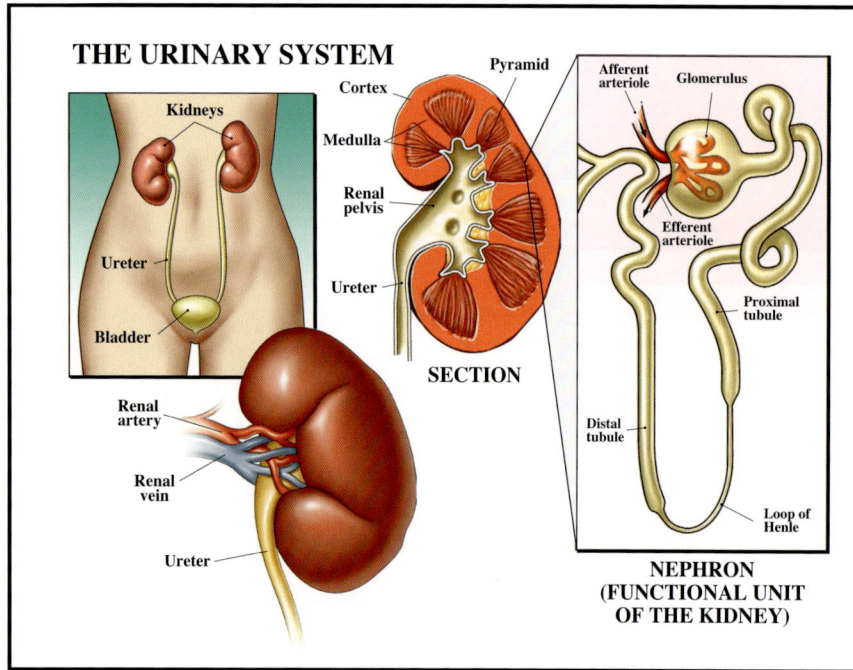

U1: The urinary system

- Waste materials are routed to the specialized capillaries of the kidneys (glomeruli) for removal from the body. Fluid, which diffuses into the renal corpuscle along with waste and metabolites, is reabsorbed through the tubule portion of the nephron, the functional unit of the kidney.

- The kidneys are located retroperitoneally in each flank, the right kidney slightly lower than the left. The internal anatomy of the kidney is arranged into pyramids (made up of tubular elements) and cortex (composed of the glomeruli). The tubules open into the calyces and renal pelvis, from which the ureter carries concentrated urine to the bladder.

- Renal disease can take many forms and can affect many other organ systems; loss of renal function requires frequent dialysis and/or renal transplantation. Total lack of renal function is incompatible with life.

ANATOMY OF THE PENIS

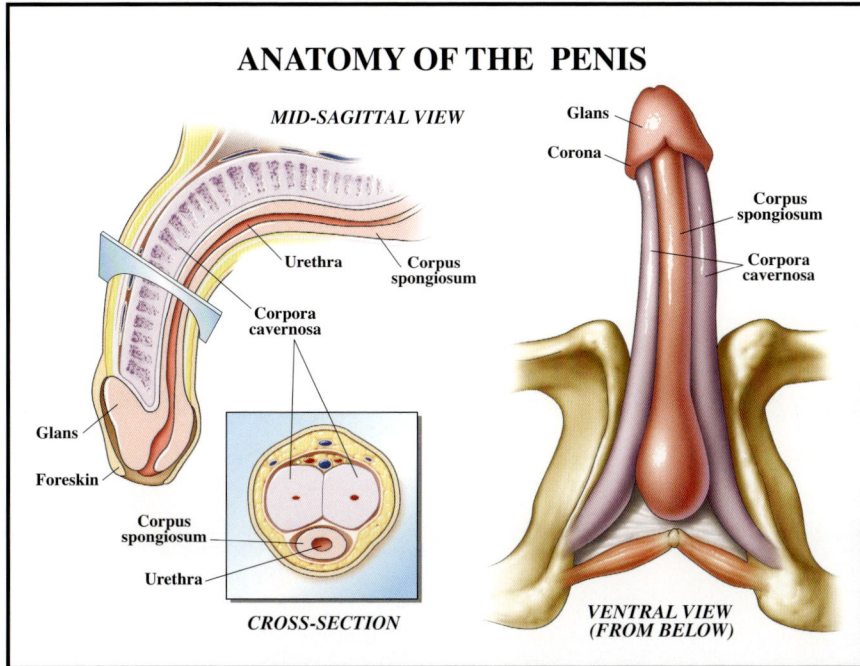

MID-SAGITTAL VIEW

Glans

Corona

Urethra

Corpus spongiosum

Corpora cavernosa

Glans

Foreskin

Corpus spongiosum

Urethra

CROSS-SECTION

Corpus spongiosum

Corpora cavernosa

VENTRAL VIEW (FROM BELOW)

U2: Anatomy of the penis

- The penis is made of three elongated masses of erectile tissue: the bilateral corpora cavernosa and the single corpus spongiosum, through which the urethra passes.

- During erection, venous blood flows into the corpus structures.

- Lifestyle and other conditions such as smoking, chronic alcohol use, atherosclerosis, hypertension, diabetes, medications, hormonal problems and psychological conditions can affect erection.

UROGENITAL

MALE PELVIC AND REPRODUCTIVE ANATOMY

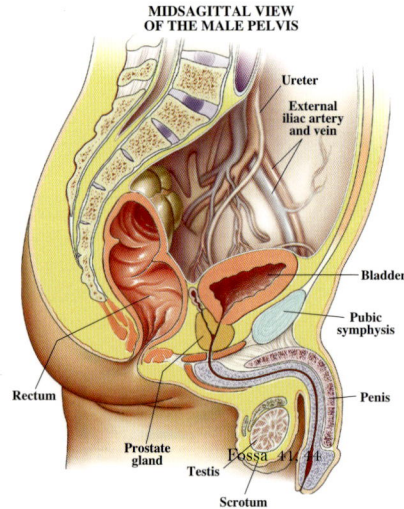

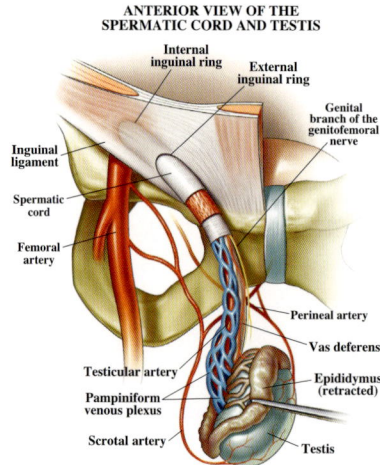

MIDSAGITTAL VIEW OF THE MALE PELVIS

- Ureter
- External iliac artery and vein
- Bladder
- Pubic symphysis
- Penis
- Rectum
- Prostate gland
- Testis
- Scrotum
- Fossa #12-14

ANTERIOR VIEW OF THE SPERMATIC CORD AND TESTIS

- Internal inguinal ring
- External inguinal ring
- Genital branch of the genitofemoral nerve
- Inguinal ligament
- Spermatic cord
- Femoral artery
- Perineal artery
- Vas deferens
- Epididymus (retracted)
- Testicular artery
- Pampiniform venous plexus
- Scrotal artery
- Testis

U3: Male pelvic anatomy

- The male pelvis contains the bladder and rectum, along with the internal portions of the reproductive system: the prostate, seminal vesicles, and the intrapelvic portions of the ductal apparatus.

- Sperm is produced in the testes, which lie in the scrotal sac outside of the pelvis. The sperm travels up the spermatic duct and is stored in the seminal vesicles. At ejaculation, sperm is released along with prostatic fluid, both of which travel down the urethra.

- The urethra has three parts: the prostatic portion; the membranous portion which passes through the urogenital diaphragm, and the penile portion.

- The spermatic cord contains a venous plexus, the spermatic artery and the spermatic duct.

INDEX

INDEX

INDEX